Keeping Chickens For Beginners
Keeping Backyard Chickens From Coops To Feeding To Care And More

Jason Johns

Visit me at www.GardeningWithJason.com for gardening tips and advice or follow me at www.YouTube.com/OwningAnAllotment for my video diary and tips. Join me on Facebook at www.Facebook.com/OwningAnAllotment.

Follow me on Instagram and Twitter as @allotmentowner for regular updates, tips and to ask your gardening questions.

If you have enjoyed this book, please leave a review on Amazon. I read each review personally and the feedback helps me to continually improve my books and provide you with more helpful books to read.

Once you have read this book, you will be offered a chance to download one of my books for free. Please turn to the back of the book to find out how to get your free book.

© 2020 Jason Johns

All rights reserved.

TABLE OF CONTENTS

Why Keep Chickens? 1

Choosing a Coop For Your Chickens 7

Other Equipment Needed to Keep Chickens 20

What to Look for in a Chicken Breed 24

Buying Your Chickens 33

All About Chicken Breeds 39

Feeding and Watering Your Chickens 55

Managing Your Chickens 67

Cleaning the Coop 84

Pests, Diseases & Predators 90

Why Chickens Stop Laying 109

Safely Storing & Using Fresh Eggs 114

Raising Chickens for Meat 116

Breeding Your Chickens 125

How To Hatch Eggs 128

Caring For Baby Chicks 134

Holidaying When You Have Chickens 140

Chickens for Exhibitions & Shows 143

Making Money with Chickens 148

Chicken Keeping Tips 153

Endnote 156

WHY KEEP CHICKENS?

Keeping chickens at home or the allotment has become increasingly popular in recent years. Partly due to health concerns, partly due to the popularity of self-sufficiency, it is something you have considered, and this book will teach you everything you need to know about buying, managing and raising chickens.

You need to understand the work and commitment involved in raising chickens before doing anything. There are many benefits from keeping chickens, and it is incredibly rewarding … the fresh eggs are just one benefit, but a lot of people don't know exactly what is involved.

Before you go any further, you need to check with your HOA, city or local council to see whether you are allowed to keep chickens. Some HOA's and cities do not permit chickens, and you can find yourself facing a penalty or fine if you keep them without permission. In general, rural areas are okay with you keeping hens. If you have an allotment, then you need to check the allotment rules as not all allotments permit chickens. Be aware that many places will permit female birds though male birds will not be allowed.

Probably the most commonly thought benefit is the fresh eggs. You may think that you are unable to tell the difference between shop bought and fresh eggs, but trust me there is a huge difference! It really surprised me when I first had a fresh egg many years ago, and now I know plenty of chicken owners that would never buy eggs from a store again.

The difference in taste is quite amazing, but very hard to describe. It has a fuller flavor, smells better when cooking and has a better all-around appearance. The yolk is a much deeper orange color. If you crack a store bought and fresh egg next to each other you will easily be able to tell the difference as the fresh egg will have a much more vibrant color.

For a lot of people, the major benefit of fresh eggs is that they know where the egg has come from and what has gone into it. People are concerned about how the chickens that produced their eggs are kept but also concerned that the egg may well be days old before it even gets to the store. Add to this the antibiotics and other chemicals that could be in the egg plus the risk of salmonella, and it is enough to put many people off store bought eggs.

The number of eggs you will get will depend on factors such as the time of year, the type of chicken and even how they are feeling! With the right breed, you can get an egg a day in peak season, though many breeds will lay 3 to 5 eggs per week. Even with just 4 or 5 hens, you could get 3 to 5 eggs a day, which is enough for many people.

Another benefit is that you can choose a breed that has an interesting colored egg shell. Plymouth Rocks, Rhode Island Reds, and Jersey Giants, to name just a few, produce eggs with brown shells. White Leghorns and Campines will produce white eggs while breeds like Araucanas and Ameraucanas produce blue and green eggs.

Chickens are also surprisingly sociable animals and genuinely like being around humans, though there are a few less social breeds. A lot of people find their chickens reduce their stress and make good companions when pottering around in the garden, though this makes it much harder to serve them up for Sunday lunch! They'll follow you around, hop up on to your lap and noticeably get excited when they see you approaching the coop.

Some people choose to raise chickens for their meat, though, this isn't for everyone. A chicken can lay eggs for anything from 3 to 7 years, depending on the breed and after that they become unproductive. At this time, some people will kill the chicken and cook it, but not everyone is comfortable doing this. If you are not, then you need to be prepared for a coop full of non-layers! It is easier to serve up your chicken if you don't get attached to it or give it a name, though you can always send it to a butcher rather than prepare the chicken yourself.

Home reared chicken meat is considered far superior to anything you can buy in a shop. The meat has better flavor, is juicier and less stringy than shop bought chicken. For anyone striving for self-sufficiency, home reared chicken meat is an important source of protein.

If you are a vegetable gardener, then chickens are a source of a very valuable fertilizer. The chicken waste is added to your compost pile and allowed to break down naturally before being used in your garden. It is a very good fertilizer, which you can buy in stores so having your own source is a good money saver!

However, there is a commitment and work involved in keeping chickens. Admittedly, it isn't a huge amount once everything is set up, but you need to be prepared for it, and it is required every single day!

About once a month you will need to change the chicken's bedding in their coop. Best to wear gloves and a mask for this so the dust doesn't get into your lungs, but this keeps your chickens healthy and disease/pest free.

Most people bring their chickens into the coop at night to protect them from potential predators and the climate. You will need a few minutes every night to herd them inside (they will learn to do this if you do it every night) and then every morning you let them back out again.

When you do this every day, you should also check their grit, food, and water. This won't take long for you do, but keeping it topped up and fresh will help your chickens stay healthy. You can also give them any fresh food as treats when you visit them.

For many people, including myself, the pros of owning chickens far outweigh the cons. The fresh eggs are exquisite, the chicken manure very useful and the chicken's good company! Although there is work involved and a big commitment, the rewards are excellent and the whole process enjoyable. It certainly makes a good conversation point with friends and family too, plus kids will love to see the chickens and get involved!

Questions to Ask Yourself Before You Start

Before you invest in chickens, you should ask yourself a few questions to determine whether it is right for you.

Are There Any Legal Restrictions?
You need to check to ensure that you are permitted to keep chickens where you live. Your city, HOA, allotment, tenancy agreement, house deeds or local bye laws could well limit your ability to keep chickens, which are classified as livestock rather than pets. Sometimes you will be permitted to keep chickens but not cockerels because of the noise they make which

disturbs your neighbors. If you are permitted to keep chickens, then best let your neighbors know purely out of courtesy.

Do You Have Enough Time?
Although chickens are not high maintenance animals, they will need visiting twice a day, regular cleaning, coop repair and more. We'll talk about this more later in the book when we discuss the routine jobs required when you own chickens.

What to Do When on Holiday?
If you go on holiday regularly, then chickens may not be for you. You will need to have someone who can look after your chickens reliably while you are away. Chickens, unfortunately, cannot be left for more than a day without needing attention. You'll learn more about this requirement later in the chapter on holidaying when you own chickens.

Can You Afford the Costs Involved?
You will need to buy (or build) a coop, a run and lots of equipment to get started, plus the chickens themselves. The initial set up is expensive, but the running costs are relatively affordable. You need to also factor in potential vet bills as well. It is important that you create a safe, healthy environment for your chickens. You can find out more about the equipment needed to keep chickens later in this book.

Have You Got the Space?
Another important consideration is whether you have enough the space for chickens. They don't need massive amounts of space, but they do need enough to roost and scratch about. Some breeds don't mind less space, but if they do not have enough, then your chickens will get sick and stop laying. This is something you'll find out more about in the section on sizing a coop.

What Do You Want from Your Chickens?
What are you hoping to achieve from your chickens? Is it eggs, meat or manure? The breed you choose will be determined by what you want to achieve. Not all good egg layers make good meat chickens and vice versa. Some chickens are very broody and great for hatching eggs, but poor layers. This is a very important thing to think about because once you have your chickens, it is a bit late to change your mind! We'll talk about this more when we discuss chicken breeds.

Risk of Predators
The danger of predators in your area is another thing to consider. If there

are predators such as foxes or domesticated predators such as cats or dogs, then you will need to provide extra protection for your chickens, whether electric fencing, strong fencing or not allowing your chickens to be free range. It is quite easy to protect your chickens from predators, but you will need to buy extra items and put in a bit more work. Putting your chickens "to bed" every night does help to protect them providing the coop itself is secure.

Phobias and Fears
This might sound a bit odd, but what if a family member has a fear of birds? It is not as uncommon as you might think. Depending on who has the phobia you may have not to own chickens or put them in a specific corner of the garden so the fear sufferer can still go into the garden. Perhaps screen the chickens off, remember you aren't going to be asking that person for help caring for your chickens!

Cockerel or Not?
Although some allotments and HOA's will allow you to keep chickens, keeping cockerels is a whole other matter. With their need to crow at ungodly hours of the morning, they are not popular with people living nearby, and may not be appreciated by your own family. A male is not needed for your chickens to lay eggs but is required if you are planning to hatch the eggs as they are needed for fertilization duties. You cannot keep more than one cockerel together as they will fight. Unless you specifically want to breed chickens, you should avoid a cockerel. If you want to hatch a few eggs, then you can buy fertilized eggs. Cockerels do make good meat birds because they are bigger than females and grow rapidly, but need extra care to prevent fighting.

CHOOSING A COOP FOR YOUR CHICKENS

This is a very important part of owning chickens and is something that you should not skimp on. It is not worth buying a cheap coop and then having to replace your chickens every few weeks as predators get inside. Buy the best you can as it will provide a safe haven for your birds and last for many years.

Whether you build one yourself or you buy a pre-made coop depends on your skills, time and finances. Whichever you decide, it needs to be solidly built, strong enough to resist predators, easy to maintain and provide plenty of room for your chickens, including space to expand your flock in the future, which is a probability!

Picture © Karl Thomas Moore

Coops come in a wide variety of shapes, sizes, materials, and qualities. You will need to choose one that is big enough for your flock and, ideally, a bit bigger as you are bound to want to get more chickens! Prices will vary widely, depending on where you are buying it from.

If you are building your own coop, then there are plenty of plans available online. You should use the best quality materials you can afford and treat the wood with a chicken friendly preservative to make it last.

Plastic Chicken Coops
These are a relatively new development and tend not to include a run, so you will need one of those too. These are well built and tend to be resistant to predators. The only issue would be size as most plastic chicken houses are on the smaller size. These are minimal work to put in place, though they do need flat ground, and cannot be repaired easily if they get damaged.

One nice thing about plastic coops is that they are much easier to clean and you can usually find feeders and water bowls that match the color of the coop. Red mite will struggle to take up residence because there is a lack of crevices for it to hide in. The plastic can easily be hosed clean and disinfected. These runs are also relatively light, meaning you can move them around your garden when necessary, but it does also mean they can blow around in high winds.

Wooden Coops
These are the most commonly found chicken houses. Typically, these are sold with a recommended number of chickens it can house. I would recommend buying one the next size up from what you want because it allows some extra room for your chickens to have space from each other, which is important. For those at the lower end of the pecking order, it gives them space to roost. Plus, it gives you additional space for extra chickens, which is bound to happen!

Picture© Josh Larios

Wooden hen houses come in a wide variety of shapes and sizes and are available in much larger sizes than plastic greenhouses. Usually, they will require self-assembly, but this isn't difficult.

Quality Matters

As with many things, in most cases the more expensive the chicken coop, the better quality it is. This applies more so with wooden greenhouses where the more expensive coops are made from better quality wood that has a longer lifetime.

For a wooden coop, you want tongue and groove boards as these interlock with each other and are much stronger. Weak boards can mean a convenient entry for a determined fox.

Check the type of wood the coop is made out of and what it has been treated with. You don't want to be painting your hen house every year to preserve it, so buy one made out of a strong wood that has a good lifespan.

Look for galvanized fittings as this will stop them from rusting. Your coop will be your biggest investment in keeping chickens, and you don't want to be replacing it unnecessarily.

The Roof

In the past, most chicken houses came with felt roofs. These are okay, but unfortunately, red mites will find their way under the felt and infect your chickens. Getting rid of the mites involves removing the roof while you kill them, which is a lot of work. If your coop does come with a felt roof, then be aware of this issue. Whichever roof you end up with, it needs to be solidly waterproof.

A corrugated steel roof overcomes the red mite problem as does tongue and groove wooden slats.

Ventilation

Your chicken coop needs to be ventilated, but it should be draft free. Although chickens don't mind the cold (they live in Canada at -20 quite happily), but they do need to sleep somewhere that is draft free. Your coop will want an inlet close to the ground and some form of ventilation higher up. The chickens should not be perched next to an opening as the draft is not good for them.

Ammonia is released from chicken dropping, which is removed by the ventilation. As the droppings heat up, so the ammonia rises. The idea is that the vent at the bottom sucks in fresh air while the ammonia flows out of the ventilation at the top, keeping the atmosphere healthy. Because of the smell of ammonia, you may not want to site your coop near your house or anywhere you enjoy sitting in your garden.

If there is too much ammonia in the coop, then this will damage the lungs and eyes of your chickens. Put your head into the greenhouse first thing in the morning when you let the chickens out and see if you can smell ammonia. If you can, then your ventilation isn't sufficient and needs improving.

Nest Boxes
Most pre-made chicken houses will come with nest boxes, which are often on the side of a chicken hours with a lid to enable to you easily access the eggs. Make sure there is a latch on this as predators, particularly foxes, will work out how to open it to get to the tasty eggs. Of course, your chickens won't always be considerate enough to lay their eggs in here, but generally they will.

If you are building your own chicken coop, then you should build a nest box on to the side that is easily accessible for you and the chickens.

Perches
These should be strong and secure, big enough for the hen to wrap her toes completely around the edges. Ideally, this is rounded. The perches need to be higher up than the nest boxes so that the hens sleep on the perches, not in the nest boxes. If they sleep in the nest box, then you will need to clean it on a daily basis as they will make them dirty.

Ease of Cleaning

This is a major influencing factor because you will be doing a lot of cleaning of the coop. It needs to be something that is easy for you to clean, remove chicken poop from and keep red mite under control. Removable parts will make life easier for you.

If the coop is difficult to clean, then it is going to make keeping chickens a chore, rather than the enjoyable hobby it should be. If you can see the assembled coop before buying, then you can determine how easy it is to clean it. If viewing a coop at a store then pull bits out, stick your head in it and understand how practical it is for you to use

Choosing the Right Hen House

Your hen house is a major investment and fundamental to your chicken keeping hobby. You want at least one square foot of floor area in the house per bird with a large access door, so you can get in to clean. Nest boxes need to be low down and in the darkest part of the house, giving the hens the privacy they need to lay undisturbed.

The perches should be higher up that the nest boxes and removable to ease of cleaning. You need a minimum of 8" of perch space per bird, though the more space you can provide, the better it is for the birds. Perches need to be about 8" apart and around 2" square.

Although plastic is a convenient material, wood is preferred because it is breathable which makes for a healthier environment for your chickens. You can find condensation a problem in plastic coops, but it isn't in wooden ones.

Look at your coop as an investment. You could buy a cheap, foreign made hen house that will last a year or two and then be worthless. Alternatively, you can spend more money, buy a high-quality one that will last 15 to 20 years and still be worth something if you choose to sell it on.

When buying your hen house, you need to consider several points, including:

- Is it strong enough to withstand the hens going in and out of it?
- How is the construction quality?
- Is the wood high quality, able to withstand the weather conditions where you live?
- Is the roof thick enough that it won't leak but also slanted enough

for the rain to run off?
- How many nooks and crannies are there for mites to hide in?
- Is the coop raised up so that rats cannot live underneath it?
- Is it solid or heavy, able to withstand high winds?
- Can it be securely fixed to the ground?
- Will it be cool enough in the heat of summer?
- How easy is it for you to clean, getting in all the nooks and crannies?
- Can you move it easily to fresh ground?
- Is it large enough for the number of birds you want?
- Is it predator proof?
- Is there sufficient ventilation?
- Is it affordable for you?

Choosing a Location

Your coop needs to be located somewhere with natural protection from the elements and some shade from the sun. Small trees planted on one side of the run can provide much-needed shade and protection from a prevailing wind. The ground should be level and not prone to flooding. Avoid locating your hen house in a frost pocket.

If your hen house is sited in one place and the chickens turn the grass into mud, then use a playground hardwood wood chip on the mud to keep your hens clean.

Free Range vs. Coops

The decision on whether to have free range chickens or to use a coop with a run attached is important.

Chickens in coops with an attached run have greater protection from predators, and it is easier for you to find eggs. Enclosed chickens are safer from disease as they do not interact with wild birds, which can carry diseases such as bird flu.

Keeping Chickens for Beginners

Although a coop needs cleaning it is easier to keep on top of rather than wandering a free range field and clearing up the mess. Your chickens will like being free range, and there will be fewer problems with overcrowding.

One advantage of free range chickens is that they will keep pests down such as slugs, snails, grasshoppers and even mice and small snakes! The downside is that they will also find any young plants, fruits, vegetables, and seedlings you grow a tasty snack too!

Free range chickens tend to be happier, and their diet makes the eggs they produce higher in omega-3 as well as vitamins A, D, E and beta-carotene. With a good-sized run, a varied diet and numbers kept under control, hens in a coop can be just as happy.

The main down side of free change chickens is they make easy targets for predators from foxes to hawks and more. Cockerels will help to protect their hens, but they are not always allowed and are not effective against larger predators. If you choose to free range your chickens in the day and put them in a barn or coop at night, then it can be hassle herding the chickens to bed if they decide they don't want to go!

Which you choose is up to you and depends on your budget, where you live and the space available to you. Most people who live in an urban or sub-urban environment will choose a coop with a run because it keeps the chickens contained, safe from predators and makes it easier to collect the eggs. In rural locations, free range is an option, but this book will assume you are a backyard chicken keeper and have a coop plus a run.

Sizing a Coop

Sizing the coop is very important if you want happy and healthy birds. If they become overcrowded then they will get ill, stop laying and even start attacking each other. Many of the potential problems associated with keeping chickens can be avoided by ensuring they have sufficient space.

Your chickens should sleep, lay eggs or shelter in the coop and then spend the day outside in their run.

Sizing your coop is relatively easy. A square meter of floor space will be enough for four large chickens or five small ones (roughly 1 square foot or 90 square cm per hen). This is a minimum, and if you can give them a bit more space, then it will make for happy, healthier hens that do not fight. This measurement must be floor space only and should not include nest boxes.

You will need around 8" of perching space per hen with the same distance between perches. For them to be comfortable and to avoid fighting, 12"/30cm is better. Although chickens tend to bundle together, the extra space means that birds lower down in the pecking order have space to perch and be comfortable. Perches need to be 2"/5cm wide with rounded edges. Larger breeds will need as much as 16"/40cm per bird to perch.

Hen houses will usually give you an idea of how many chickens it will hold, but this always tends to be less than generous. If a coop states it can house five beds then keep no more than four in it, so they have enough space. Ultimately, you need to give your chickens as much space as you can because it will result in healthier, happier hens. Remember that larger breeds such as Orpingtons will need half as much space again to be comfortable.

Nest boxes should be lower down than perches to prevent the birds roosting in them. Fill them with straw (avoid hay as it will contain mold spores). You want a minimum of two nest boxes, providing one nest box for every four hens.

At the end of the day, buy the biggest hen house you can afford and fit into your desired location. This ensures your hens are comfortable and gives you room to expand your flock without the expense of buying a new coop.

Cages and Runs

The next most expensive item on your shopping list is a chicken run and will be the weak point in your predator defense network.

There is no legal guidance for space for a hen so you must use your common sense here. As a backyard keeper, you will want your chickens healthy and happy, so they are productive, so it is in your best interests to provide plenty of space.

Picture © ChickenMan

Think about how many chickens you have and whether there is enough room for them to walk about, peck, drink, scratch and get away from each other in the case of disagreements.

Your choice is to save time and buy a pre-made run or save money and make your own. Whichever you choose, you need to ensure it is predator proof!

When you buy a chicken coop, it may well come with a run, which is likely to be undersized. You can use this one for when the weather is bad as they won't want to be outside so much. You can then buy or build a larger run for use when the weather is better. If you have the space, then two runs are better because you can alternate which one your chickens use.

If you grow vegetables or want your garden to stay looking good, then you will need a run. Although chickens control pests, they will strip your vegetable plants of fruits and eat all your seedlings! Many a gardener has gotten excited about using chickens for slug control only to find out they've eaten all the seedlings or destroyed their crop! Of course, you can fence off areas you don't want them to get to if you do want them to run free, but some breeds will still find a way to get over them.

An alternative is to use a portable run and move your chickens around the garden during the day. This stops them ruining the ground and gives them fresh grass. You will need to move them every 3 or 4 days to prevent them permanently damaging the grass.

Look at the quality of wire on the run. You want small rectangular wire or square mesh wire because it is harder for predators to get their teeth in it. Cheap, thin wire won't provide enough protection as it is far too easy for predators to break through with their teeth.

You want latches and bolts that are securely fixed and galvanized, so they do not rust. Look for stainless steel screws too so they do not rust and weaken.

Some predators, such as foxes, will attempt to dig under the run. You can dig wire into the ground around the outside of the run to stop this, or you can put a wire mesh bottom on to the run which stops the fox getting inside, even if it does dig down under it.

If you are building your own run, then make sure that the wire you buy is thick enough that a predator cannot tear it. If you are in any doubt, a double layer of wire around the bottom half of the fence will provide extra security. Pay particular attention to any joins between sections of wire and where the wire joins the wood. This needs regularly checking for any signs of wear and tear.

Foxes can run up fences, so your chicken run will need to either have a wire roof or be at least six feet tall and sloping outwards at the top. Gates need to be secure and cannot be pushed in at the corners; foxes are very sneaky indeed and will find any way they can to get at their prey.

A good way of building your run is to make separate panels and then screw them together. This is an easy way to make the run and you can secure it firmly.

If you do want to build your own run, then there are plenty of designs online that you can follow or use as inspiration for your own design.

Chicken Fencing

Chicken netting or rabbit wire is a good choice for a fence. This needs to be, as mentioned previously, 6 feet tall, higher if you have bigger predators than foxes in your area. It should also be buried 12-15 inches into the ground to prevent predators digging under your fence. Depending on your soil type, this can be hard work but is essential because otherwise, predators such as foxes will dig under it. Fix wood to the bottom of the wire where you are burying it to keep it rigid and prevent the wire being moved. In addition to burying the wire, you can put boards around the bottom of the fence for extra security. This keeps the fencing rigid and stops any soil, wood chip or sand from in the run escaping.

Electric fencing is a good option, and more on that in a moment, but good old fashioned wire fencing will work very well. If you build the fencing securely, so there are no weak spots there will be no way for predators to get in or your chickens to get out.

If your chickens roam free, then you can use chicken netting to keep them off your flowers and vegetable plants. This isn't recommended for a run as it isn't predator proof.

Electric Fencing for Your Chickens

If predators are a serious problem in your area, then you should consider electric fencing. If set up properly then it is safe, so you need to make sure it is installed correctly. It consists of a high voltage generator that sends a high voltage once a second down a wire which shocks the predators (or you or your chickens) if you touch the live wire.

This short shock isn't dangerous but is unpleasant. Pregnant women should avoid electric fences as there is a chance it could harm the unborn baby. Both the chickens and predators will very quickly learn to keep away from the fence because it is unpleasant for them. You do need to make sure that non-predators such as hedgehogs cannot get trapped in the wire.

An electric fence energiser will tell you how many Kilometres they can power plus you will need a good quality, earth rod. You will also need insulators to hold the wire to fence posts. The lowest wire should be about 6 inches off the ground, so it doesn't get shorted out by grass but is low enough to deter digging. Another wire needs to be knee height, which is about nose height for a fox and acts as a great deterrent. The third wire should be at the top of the fence which will stop animals climbing. Remember to regularly strim the grass around the edge of the fence to stop the grass from touching the electric fence, ensuring the wire remains visible.

Electric poultry netting is another option, particularly as a temporary fence if you want to move your chickens around. Remember that because the netting has multiple strands, typically 12, the distance to electrify quickly adds up. A 25m square run with 12 strand chicken netting has a 100m perimeter, which doesn't sound too bad. However, as there are 12 strands, your perimeter distance is actually 1.2km. Take this into account when sizing your energizer otherwise your fence will have weak spots and won't provide an effective barrier.

You can buy kits for both electric fencing and netting, online and from farming stores. One very important consideration is warning notices. Even if your chicken run is in your garden, you should still put notices up to protect visitors, family and, oddly enough, intruders! Ensure they are clearly visible to both adults and children, remembering that children, in particular, will be drawn to look at the chickens and can easily touch the electric wires without thinking about it.

Electric fences are a very good way of protecting your chickens and keeping predators out. So long as they are properly installed and regularly maintained they are good. Yes, they are not cheap, and you have the ongoing running costs to consider, but if you have a serious problem with predators, this is the best solution to keep your chickens safe.

Other Equipment Needed to Keep Chickens

The chicken coop and run are the major costs involved with keeping chickens, but there are several other items you must buy, some of which are ongoing costs. This doesn't include any of the equipment you need to hatch and rear chicks.

Feeder and Waterer
The best types of feeder and water container to get are those that are either suspended off the ground or hang. This stops the chickens jumping on to them and pooping in their food or water. You can buy ones that automatically refill, which allows you to go away for a few days if necessary. These aren't too expensive but are vital. Galvanised metal is much better than plastic as they will last longer and are less likely to be damaged.

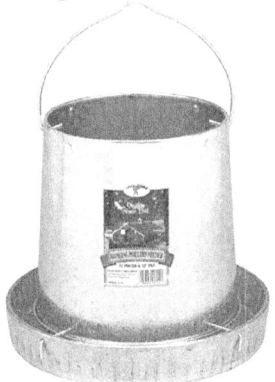

Galvanized metal hanging feeder, available on Amazon.

Food

A complete food with the correct mixture of vitamins and minerals is the best option to go for. These can be found at some pet stores, online, at farming supply shops or some garden supply stores. It is worth buying in bulk and storing somewhere that rats and mice can't get to.

For chickens, under 20 weeks old you will need a starter or developer feed, though read the label on the food as some are only suitable for younger chickens. When they start laying or are over 20 weeks old, they will need a complete "layer" food. Meat chickens benefit from a "grower's" food, which helps to grow fast and fatten up.

Scratch

This isn't essential but the chickens love it, so it is strongly recommended. This is a mixture of diffcrent grains such as oats, wheat, rye, and corn. It is usually kept, securely, in the run and some given to the chickens every day as a treat. It is thrown on the ground, and the chickens go crazy scratching around and pecking at it. Remember this is a treat and not their staple diet. If the egg shells become weak and break easily, then you are giving them too much scratch and need to cut back, so they eat their regular food and get the right levels of nutrition.

Grit

Grit consists of very small rocks that the chickens store in their crop. This breaks down the food before it enters their stomach. As chickens do not have teeth, grit is vital because otherwise, your chickens cannot digest their food properly. You will usually keep this in a small container near their food or mix a little with their food.

Oyster Shells (Calcium)

This is an optional treat that you can offer your chickens, which isn't totally necessary if they have a complete food. However, it isn't expensive and lasts a long time so is useful to offer your chickens. If your hens spend a lot of time foraging, they may not get enough calcium from their food leading to oddly shaped eggs, slow laying and even egg eating! This is a vital source of calcium for your chickens and is worth including even if they eat a complete food.

Bedding

Your chickens will appreciate a good quality bedding. It makes the coop floor softer for them to walk on, absorbs the droppings and helps to stop the smell. The soft bedding in a nest box protects the eggs, so they don't crack when they hit the hard floor.

There are many different types of bedding you can choose. One of the best available is pine shavings, not pine chips. Other materials do not absorb the waste material as well and can encourage pest infestation. You need to bedding to be a minimum of 2" (5cm) thick.

Dust Bath

Dust baths are very important to chickens. In the wild they will dig a shallow hole, loosen the earth up and then cover themselves in the dirt, shaking it off later. This is vital for them because it helps stop lice and mites taking hold.

Picture © Powerfox

It is worth providing your chickens with an artificial dust bath, which is simply a box filled with about six inches of dusting powder. You can buy this or make it from one part each of road dust, sand, diatomaceous earth and fireplace ashes.

Torch

Best to get one with adjustable beam strength and ideally a head torch. This is very useful for going into the coop when the chickens are roosting and are easier to handle.

Scissors

A strong, sharp pair is required for varying cutting jobs including clipping flight feather.

Toenail Clippers / Nail File

In most cases, your chickens will naturally keep their toenails worn down. However, if the ground is soft for an extended period of time you may need to help them trim their nails. Sometimes cockerels will need their spurs trimming.

Leg Rings
These are useful for identifying your flock in case any escape but are also very helpful if you are treating an individual bird or the whole flock and want to mark birds that have or have not received treatments. A pair of wire cutters is useful for removing these.

Petroleum Jelly
Useful as a lubricant on catches and locks as well as to put on the combs of birds to prevent frost bite in cold weather. It can also be applied to dry patches of skin and is useful in treating scaly leg mite.

Pet Carrier
Necessary for transporting or quarantining chickens. Plastic dog or cat carriers are suitable for single chickens, though larger dog cages are needed for multiple chickens. Remember to disinfect the cages after every use to remove the risk of spreading infection.

What to Look for in a Chicken Breed

Choosing the right chicken breed is very important because choosing the wrong one means you have to get rid of your flock, which may not be easy. There are a huge variety of breeds, and they are different in the way they look, their appearance, their laying ability and more. There isn't one breed considered to be "the best, " but there are breeds which are better suited for your specific needs. In this section, you will get some help choosing the best breed for your individual requirements.

Before you start, think about what you want your chickens for …

- Meat
- Eggs
- Companionship
- Exhibitions
- Their looks
- For a small space
- For a warm climate
- For a cold climate

These are the first considerations to make because some breeds are better suited for certain situations and needs than others.

Your flock may not consist of just one breed. You may have a variety, some good layers, some good meat chickens and then your chickens for exhibitions and shows. Many people will keep a mixed flock, though be aware that you will need to separate them if you plan to breed pure bred chickens.

If you want to show your chickens at any time, then you are better getting your chicks from a breeder who already shows chickens. They are more likely to conform to show standards and be higher quality. Although chickens from most hatcheries are good quality and healthy, they are not up to the breed standards required for shows. These special chickens are more expensive than hatchery chickens but are much more likely to be of show quality.

Of course, which chickens you get will also depend on availability. While all of these breeds are available they may not be local to you or in your country. You can order fertilized eggs mail order and hatch them yourselves if you want a rarer breed but to get baby chicks, you will have to travel unless you are fortunate enough to be able to buy them locally.

Dual Purpose Breeds

These breeds are great for the backyard flock, being good for both meat and eggs. They tend to have a good temperament, are capable of foraging and are also good brooders. These are ideal for people who are not sure what they want their chickens for or for people that don't have a lot of space but want chickens good for a variety of purposes.

Some examples of dual purpose breeds include:

- Australorp
- New Hampshire Red
- Orpington
- Plymouth Rock
- Rhode Island Red
- Sussex
- Wyandotte

Egg Producers

If you are purely after eggs from your chickens, then you want a breed that is good at converting their food into eggs and lay regularly. Laying breeds tend to be slender, so are not very good for meat. They are suitable for backyard flocks but tend to be a bit nervous and good fliers, so put a roof on their cage! Laying breeds tend to be less broody than other breeds.

Examples of egg producing breeds include:

- Black Star
- Campine
- Dominique
- Fayoumi
- Golden Comet
- Hamburg
- Leghorn
- Red Star

Meat Producers

Meat producing breeds tend to be larger birds with big thighs and breasts. They are not usually very good layers because they don't waste their energy on eggs. Most of the big roasting chickens you buy at the supermarket are a Cornish-Rock cross hybrid that is, by far, the best chicken for meat production.

Popular varieties include:

- Brahma
- Cornish

- Cornish-Rock Cross
- Freedom Ranger
- Jersey Giant
- Red Ranger

Standard Size Chickens

Space is important for healthy and happy chickens, which means you may not be able to buy larger breeds of chicken. Standard breeds will typically weigh between 4 and 8 pounds and are the size we are all used to seeing.

These are the types of chicken you are most used to and will usually produce standard sized eggs. There are some breeds which are large breeds and need more space/food.

Bantam Chickens

These are miniature chickens, around a quarter of the size of standard breeds. There are often bantam versions of the standard breeds, e.g. a bantam Brahma. There are some bantams, such as the Serama and Dutch which do not have larger equivalents and are purely bantam sized hens.

Bantam breeds will lay eggs, but they are about half the size of eggs from standard breeds. However, they do lay fewer eggs than their bigger counterparts so are not suited if you are specifically after eggs.

You are forgiven for wondering why anyone would want a chicken this size, though they are very cute and cuddly!

Bantam chickens are very useful for anyone who either has very little available space for their chickens or need to keep their chickens in the coop all the time. They are small and have a very calm temperament, in general, which makes them very good for children to be involved in their care.

Bantams often make good show birds, but they are very broody. Some people will keep bantams such as Silkies or Cochins to hatch eggs from less broody, larger birds.

Some of the smallest bantam breeds are:

- Serama – less than 1lb in weight
- Nankin – around 1½lb
- Dutch – around 1¼lb
- Bearded D'Uccle – around 1½lb
- Bearded D'Anver – around 1½lb
- Booted Bantam – around 1½lb

Egg Shell Color

For some people, the color of the egg shell is important for whatever reason. The color of the shell does not have any effect on the flavor, so you can keep a variety of breeds and have a great selection of colored eggs.

Colored shells are produced by the following breeds:

- Brown – a lot of breeds including Plymouth Rock, Brahma, Rhode Island Red

- Dark Brown – Penedescenca, Welsummer
- White – usually from slender egg layers such as Leghorns, Hamburgs and Campines
- Creamy/pink – Salmon Faverolles, Cochins, Old English Game Bantams, Dorking
- Blue – Easter Eggers, Cream Legbar, Araucanas
- Olive Green – Olive Eggers

Plumage Choices

Chickens come in a huge variety of colors and patterns. They are very pretty birds so you can choose birds based on the fact you like the color, which some people do!

If your birds are going to be free ranging, then choosing colors that blend in with the landscape to help protect them from predators. White plumage is the easiest for predators to spot, but for meat chickens, it is advantageous as plucking doesn't leave dark spots.

Silkies and Polish chickens have head feathers which can get in the way of their vision, making them particularly vulnerable to aerial predators. This makes these breeds unsuitable for free ranging.

Feathered legs look good on a chicken and are advantageous in colder climates, but it does mean the chickens are more prone to scaly leg mites.

Climate Considerations

Most chickens are pretty tolerant of a wide range of climates, but if you have very cold winters or hot summers in your area, then you may want specific breeds that are best adapted to that climate.

The comb of a chicken is very susceptible to frost bite. Petroleum jelly can protect the comb, but breeds such as the Wyandotte which has a rose comb is less vulnerable to frost bite than breeds with a single bladed comb. Large flat combs and wattles are less susceptible to frost bite than small wattles and thicker combs.

Chickens that have rounder bodies will lose heat slower than those with slender bodies. Therefore, the egg laying breeds which have more slender bodies are not suitable for very cold climates as they feel the cold more.

Some of the breeds which have round bodies and are labeled as cold hardy have single blade combs, which are very susceptible to frost bite. These tend to be older breeds from the days before people realized chickens felt pain, and frost bite is very painful!

Leg feathers help insulate the chickens from cold but check carefully and regularly for scaly leg mite. Fuller feathers help provide better insulation.

Popular cold hardy and frost bite resistant chicken breeds include:

- Brahma (males are susceptible to frost bite on their large wattles)

- Buckeye
- Chantecler
- Wyandotte (males are susceptible to frost bite on their large wattles)

Heat tolerant hens have opposite characteristics to the above. They tend to be slender breeds with single blade combs, losing heat quickly, so they stay cool. Breeds well suited for a hot climate include:

- Andalusians
- Black Faced White Spanish
- Campines
- Catalana
- Lagenvelder
- Leghorns
- Naked Neck

Temperament

Chickens have personalities, with certain breeds have certain temperaments and behavioral traits bred into them. The meat breeds and dual purpose breeds have a much more relaxed temperament than the nervier slender layers. Some breeds, such as Silkies and Cochins will go very broody whereas dual purpose breeds tend to be very good foragers.

Breeds with a pleasant, calm temperament are:

- Australorp
- Brahma
- Cochin
- Faverolles
- Orpingtons
- Plymouth Rock
- Success
- Wyandotte

There is no need to introduce aggressive breeds into your flock as they can end up causing trouble and harming the other birds.

The original cock fighting breeds are much more feisty than many of the chickens bred for meat and eggs. These breeds, which are now usually used for showing include:

- American Game Bantam
- Cubalaya
- Modern Game
- Old English Game

These will need separating from the rest of your flock as they could bully other breeds of chicken and damage them.

Of course, all the above information provides general guidelines for the traits of breeds. There is variation within the breeds and while say a Leghorn will typically lay a lot of eggs; you may get one that doesn't. So, do expect some variations, but in general, if you go for a breed with specific traits, you will get those traits with the occasional exception.

Buying Your Chickens

Once you know what you want your chickens for, you will identify a breed or breeds that have those qualities and is suitable for your climate. Now it is almost time to buy your chickens. Before you do though, you need to make sure that their new home is set up and ready for them to take up residence, plus you have all the necessary equipment for feeding and looking after them.

If this is the case, then you are ready to buy your chicks.

In my experience, it is better to buy from a small, local breeder as chickens are a hobby for them, so they take good care of the birds and are more likely to give you good quality chickens. Hatcheries are an option, but these provide chickens on a more industrial scale, and there can be more of a risk of getting unwell chickens.

Another alternative is to get ex-battery hens. These are usually very affordable and will have a few years laying left in them. If you search online, you should be able to find somewhere rehoming them close to you.

You should see the chicks and pick the ones you want directly, going home with them in your cardboard box or cat carrier. Look for the healthiest ones you can see. Avoid birds that wheeze, are sleepy, look like they are fluffed up against the cold, hold their mouths open or have bubbles or foam in the corners of their eyes. A bird that is healthy will be active and alert, moving around to forage for food.

You can feel under pressure to buy if you go to look at birds, so it is often better to tell them you are coming to look with a view to purchasing shortly. If, when you see the birds, you are not happy with their health or appearance you should buy them elsewhere otherwise you could end up with sick chickens that don't live long, or worse, infect your current flock. We'll talk more about introducing new chickens into an existing flock later.

When you go to buy chickens always take a carrier of some form with you. A cat carrier will work or a cardboard box with deep sides. The breeder will appreciate this, and it will mean you can take your chickens home the same day.

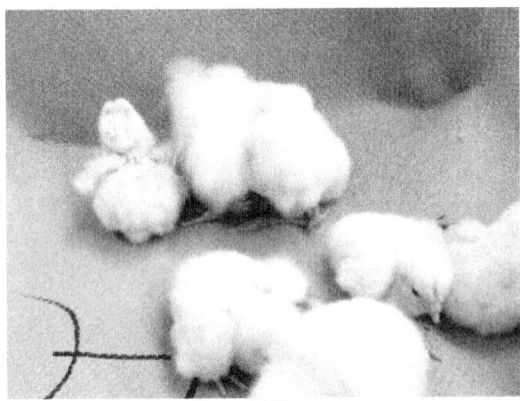

Picture © Creigpat

There are a number of places you can buy from, which I'll list below, but be aware that cheap chickens are not necessarily the best. If you want high quality, healthy chickens then expect to pay a premium price, particularly if you are after a rare breed.

- Breeders – buying pure breeds from people who show their birds is a great way of getting really good chickens. Yes, they will be expensive, but you know you are getting top quality. This isn't for everyone, and most people will prefer a cheaper option, but if you

want to show your birds at any time, this is the best place to buy your chickens from.
- Poultry Shows – a good place to get a wide variety of birds from and you can ask questions of breeders and experts there. You are very unlikely to see any sick birds, and you can often find some of the more unusual breeds at these shows. Country or county shows may well have chickens for sale too.
- Private Breeders – these are people who hatch eggs at their home, perhaps it is a hobby, business or extra income for them. This is a good place to get chickens from, though go and see the birds before buying them. You can often find adverts online on poultry websites.
- Word of Mouth – if you know someone who has chickens you can ask them where they got theirs from. They may well be able to direct you to someone locally who can provide good quality chickens.
- Poultry Auctions – these can be touch and go for quality, so buyer beware! Although you can often find reasonable quality chickens at auctions, a lot of people use them as an opportunity to get rid of their substandard chickens that they cannot sell elsewhere.
- Hatching Your Own – an option for some people now that incubators are cheaper and easier to use. You can find fertile eggs for sale online at a wide variety of sites, including eBay. This can be a great way to get the more unusual breeds or to get chickens if you do not live close to a breeder. You'll learn more about hatching eggs in a later chapter.

Choosing Healthy Chickens

When you buy chicks, it is best to choose your birds and leave with them the same day. This ensures you get the chicks you wanted and know are healthy. Coming back to collect them a few days later gives the less scrupulous breeder the opportunity to swap healthy chickens for unhealthy ones.

A healthy chicken is going to be active during the day. It will be curious, moving around and looking around for food. A chicken that is drowsy, sleeping during the day or not moving much has something wrong with it and is worth avoiding.

Chickens are very susceptible to stress, which is the major cause of disease in chickens. Moving chickens to a new environment causes stress and if the chicken is unwell to start with it can prove too much for them.

This is one reason why it is important that you choose healthy chicks.

Look at the combs on the hens. They should be a bright red color. If they look blue, and particularly with cockerels, then this indicates circulation problems or potential diseases/organ problems. If the cockerel has scabs on its comb, then it means it's been fighting, which will cause you to question how the breeder manages their chickens.

You should always inspect the birds before you buy them. If the breeder gets edgy about this and starts making excuses, then you should leave as this is a major red flag. Inspecting them lets you check them over. Look for lice, which look like small grains of skin colored rice, on the feathers and skin. They will crawl away from the light and move pretty quickly. Although lice are easily treated, it could be an indication of the general health of the chickens.

Look at the chicken's vent and check the feathers; they should be clean. If there is a lot of muck around this area, then it can mean the chick has worms.

Check for missing feathers as this can indicate bullying in the flock or that the hen isn't in the best condition. If the bird is molting, then you don't want to take them as molting is a stressful time for chickens and moving them now can make them unwell.

On the legs look out for scales sticking out from the legs as this indicates scaly leg mite, which is uncomfortable for the chickens. Again, this can be treated but does raise questions about the general health of the flock.

Both the upper and lower parts of the beak should meet and should not be crossed over. Check the toes are straight and not bent. Although these don't affect the bird's ability to produce eggs or meat, it does indicate a genetic problem, so avoid breeding birds with these deformities.

Make sure there are no bubbles in the corners of the eyes and that the eyes are clean. When you pick up the chick, it should be breathing clearly with no wheezing or coughing noises. If you are unsure what you can hear, hold the chick up to your ear and listen to it breathing. If it does have respiratory problems, then you should avoid the chicks.

Buying your chickens should be a fun process. If you have any doubts about the breeder or the quality of their chickens, then go elsewhere. It is better that you get healthy chickens as you will save on vet bills and have birds that will be productive throughout their life time.

Chicks or Eggs?

Should you buy baby chicks or hatch your own?

This is very much a personal decision with there being no right or wrong answer. If you want to buy an incubator and see chickens hatch, then buy eggs. If you have children, this can be a really exciting part of the process plus a great way to get them interesting and involved in the chicken keeping experience.

Buying chicks mean your birds are going to start laying quick plus you know you are not getting any cockerels as they will have already been sexed by the time you buy them. If you get really young chicks from a local breeder, then they will often swap out cockerels with female birds when they are old enough to be sexed.

Buying eggs is a bit of a lucky dip as you could end up with a cockerel or cockerels, and if you cannot keep them, you must either dispose of them or rehome them. Sexing a chicken isn't easy to do when they are young, and you may have to either wait until they grow older or hire a professional!

There is a job of chicken sexer, and it requires a great deal of training to perform plus it pays extremely well!

Beginners will generally want to buy chicks because it is easier and they know what they are getting. As you grow in experience, you may want to try hatching eggs. If you want a particularly unusual breed, you may have to go for eggs. If you do choose eggs, be aware that they are slightly more likely to hatch males and you may well have to deal with these unwanted males, which is not something everyone wants to do.

All About Chicken Breeds

There are hundreds of breeds of chickens, and I don't need to bore you by listing every single breed. You will learn about some of the more popular breeds for the backyard farmer in this chapter. You will also learn a bit more about some of the most popular egg laying and meat chickens which we have already mentioned.

If you know what you are looking for in a bird, then it is easy to determine a breed, but if you are not sure then this information may help you decide upon the right breed for you.

Pure breed chickens are more expensive, but, like a pedigree dog, have all the traits of that breed. Hybrids are usually more robust and often more docile, making them better for new chicken owners.

Araucana
This breed originated in Spain and made its way to South America with the Spanish. They are cute chickens with tufts of feathers covering their ears, a Tudor ruff around their neck and come in a wide variety of colors. The eggs are a solid blue with strong shells and are relatively big compared to the size of the bird. You can expect between 170 and 200 eggs per year from each bird. This breed comes as a standard sized bird and is also available as the smaller bantam.

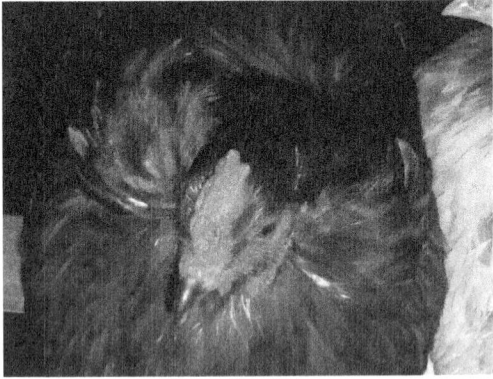

Picture © LeGrkery

Barnevelder
This bird is from Holland and is known for being a friendly bird that is easy to care for. It is a heavy bird and a poor flyer, though very good looking as a silver or gold lace. It is available as a large bird or a bantam, laying dark brown, often speckled eggs. Typically, you will get between 180 and 200 eggs a year. This breed is particularly vulnerable to Marek's disease, so make sure any hens you buy are vaccinated (done at one day and two weeks old).

Picture © Paul Pleece

Brahma
This is a very large bird and so not suitable for keeping indoors or in small spaces, although there is a bantam variety. Coming from India, they are were originally referred to as Grey Chittagongs and are known to be calm birds. They are a very good looking bird with a feathered skirt and feathered feet. They are vulnerable to scaly leg mite and do not fly, making them easy to handle and great for children. They are good broody hens. You will

typically get around 120 medium sized eggs per year from one hen.

Picture © Rillke

Cochin
Another ornamental bird that looks great and has feathered feet. It doesn't fly well and is a very good brooder. Originally bred for both meat and eggs, you can expect between 150 and 200 medium sized eggs every year.

Picture © Hagen Graebner

Dorking
Another dual-purpose bird, good for both meat and eggs. It's unusual in that it has five toed feet and is available as a bantam and a normal sized bird. It is very susceptible to frost damage. The cockerels grow to be very large indeed, though the females are good mothers. They will produce between 150 and 200 eggs every year.

Picture © 4028mdk09

Dutch Bantam

Only available as a bantam, this is a fascinating bird being very easy to tame and breed well. The male bird looks just like a miniature larger cockerel and is known to be noisy and more than a little feisty, though because of its size it doesn't cause a lot of damage. The females produce small white eggs and are good mothers.

Picture © stephen jones

Black Rock

This is a docile and hardy chicken that has been bred to be a free range bird. It has some mite resistance and is well insulated against cold weather. A long-lived breed it has one of the longest egg laying periods and is a prolific layer, producing over 280 eggs a year! It is a distinctive looking bird with copper neck markings, though it enjoys scratching and digging.

Bluebell

This is a docile bird that is very friendly, perfect for a free range environment. It is a very pretty bird and will lay approximately 240 eggs a year that are large and dark brown in color.

Calder Ranger

Another good bird for a free range environment it is a prolific egg producer and can produce 300 to 310 eggs a year!

Cream Legbar

This is a nosy bird that makes good mothers, whether for their own eggs or others. Interestingly, you can sex the chicks by their head color! Male birds will have rougher spots of color on their heads which spread across their body as they grow. The hens will produce upwards of 180 blue to green eggs a year.

Picture © Jack Berry

Faverolles

This breed originated in northern France and were bred for egg production as well as meat. These are calm birds that are happy to be handled, but they can get bullied by other, more feisty breeds. They are best suited to a single breed flock or in a flock of birds of similar temperament. They are not particularly good layers, despite their breeding, and you can expect around 100 small eggs a year.

Picture © stephen jones

Hamburgs

An old breed, originating in Germany around the 14th century, this is a good looking bird. This breed is well known as a consistent layer throughout the year with a long laying life time, though they do not lay a lot of eggs every year. They are a smaller breed, with hens weighing on average about 4lbs. Hamburgs do not like being confined, preferring a free range environment and are very curious, loving to forage and explore. They are not a broody bird.

Leghorns

Leghorns are one of the top egg producing breeds, though they are not cold hardy because of their large combs. This is an excitable breed that is noisy and quite shy around humans. They are a very active breed and like to move around a lot, meaning they need a good size run or free range field. They do not like confinement because of their need to be active. You can expect around 300 or so eggs every year, with them still producing eggs during winter and for as much as ten years. This is not a broody breed.

Picture © Dejungen

Marans

Originally a French breed, pure-bred Marans are hard to find as they are usually cross bred with other breeds. They are docile, though enjoy foraging and have good disease resistance. This breed is available as both a large bird and a bantam. You can expect a minimum of 150 to 200 eggs every year, which are large and dark brown in color.

Picture © seppingsR

Marsh Daisy

Originating in the UK, this is a slow growing bird that is hardy. It forages well and is suited to a free range environment. Generally calm, the bird flies well and is striking to look at with green legs, attractive plumage, and a rose comb. It produces small eggs, and you can expect between 100 and 150 per year.

Orpington

A popular bird that is tame and easy to handle, this is a broody bird. It has soft feathers and does not enjoy damp or muddy environments. This breed lays throughout the winter and will produce between 170 and 200 light brown eggs every year. It is worth knowing that the female birds will feather before the males, which makes them a little easier to sex.

Old English Game

This is a very noisy and active bird that isn't well suited to the beginner. They were originally bred for cock fighting and are not tolerant of other birds. Cockerels should never be kept together because they will fight to the death. This breed is very hardy, but it does not like being confined. They fly well and will roost in trees if you let them. The females are good brooders, and the bird is available in a variety of colors and as a bantam breed too. Typically, you will get between 100 and 150 eggs a year with a cream tint.

Picture © 4028mdk09

Pekin

An ornamental bantam, this cute ball of feathers will be a huge hit with your children! They love being held and are very sociable when they think they are getting treats. This breed has a long feather skirt, so is not suited to damp or muddy environments. Because of their size, they will need somewhere to dry. They are not big layers, producing around 95 small eggs a year. However, this breed is very broody and make excellent mothers, being very good at hatching eggs from other chickens.

Picture © Camille Gillet

Polish

A breed originating in Europe it is very easily distinguished by its feather top hat. It is available as both a large fowl and bantam. Because of the "hat", the bird is more susceptible to lice and is quite timid, due to limited vision. They produce 3 to 5 medium sized eggs a week and will stop laying in winter. They do not like damp weather and are not a broody breed. They are very submissive and should not be mixed with other breeds as their crest will be pecked and they are often bullied.

Rhode Island Red

This is a very popular breed, and rightly so. It is a prolific layer, producing 250 to 300 brown eggs every year. They are a hardy breed, being bred to withstand the cold New England winters. The birds love to free range and are available as both a bantam and a large fowl. They are short tempered birds, so not well suited to be mixed with more submissive breeds. They are friendly to humans and quite easy to tame. This is an ideal bird to start off with.

Silkie

This is a cute bird, available as both a large fowl and a bantam. They are notable for their fluffy, not feathery, plumage. They have five toes, blue earlobes, and a dark colored skin. They are a very docile breed and are excellent brooders. This breed, though, is not hardy and really doesn't like cold or damp weather. They need to be fussed over and looked after and will be bullied by other chickens. You can expect about 100 medium to small sized, cream colored eggs every year, though broodiness often interrupts laying.

Picture © Aaron Jacobs

Speckledy

A hybrid of the Maran and Rhode Island Red this is a free ranging bird that is a great forager. These birds are very sociable and are not aggressive. They will lay 260 or so dark brown eggs per year and also make good foster mothers.

Sussex

Available as both a bantam and large fowl this breed comes in a lot of different colors. It is a docile breed that is very adaptable. They can live well in confinement but are great foragers so good as free range birds. Sussex chickens love humans and are very tame. The males are big and very aggressive while the females are excellent mothers due to their broodiness. You can expect somewhere between 240 and 260 medium to large brown eggs every year.

Picture © Alpha

Welsummer

Coming from the Netherlands, this breed is available as either a bantam or a large fowl. It likes to free range and is a good forager, though can be kept confined. It is a friendly bird that doesn't mind being handled. Although the females are broody, they are not good mothers. The baby chicks auto sex, making it very easy for you to pick the female chicks. You can expect about 160 large eggs per year.

Picture © Josh Larios

Welsummer on left, Delaware in middle, Buff Orpington at rear

White Star
Also from Holland, this is a small bird that is a frugal eater. It isn't a long-lived breed, putting its energy purely into egg laying. You can expect up to 320 white eggs every year.

Wyandotte
Bred in the United States this breed comes as both a large fowl and a bantam. It is happy in a run or free ranging and is quite a vocal bird. It is also extremely friendly, making it great for beginners or children. The females make good mothers and will turn broody after laying eggs. Watch them closely at this time as they can stop eating because they do not go outside. Eggs are tan or pale brown in color, and they will lay between 200 and 240 eggs every year.

Picture © ripperda

Best Egg Laying Breeds

Eggs are one of the main reasons people keep chickens, and there is nothing like the feeling of collecting fresh eggs. While you have learned about some of the many breeds, and now we will focus on what most people consider to be the ten best egg laying breeds. Getting the right breed is vital if you want a regular supply of eggs throughout the year.

- Hybrids – these are specifically bred for egg production, with one of the most common being the Golden Comet. This breed has a relatively small appetite but will produce around 280 medium, brown eggs every year. These breeds are not known for their broodiness and will usually produce some eggs through winter.
- Rhode Island Red – A very popular breed these will produce 250 to 300 eggs every year that are of a medium size and brown in color. Being very friendly birds, these are a great choice for the first time chicken owner.
- Leghorn – Another popular breed producing white, medium sized eggs. You can expect around 300 eggs every year. This breed is good for a beginner, but they are shy and known to be difficult to tame.
- Sussex – A great breed that will lay somewhere in the region of 250 brown to creamy white eggs every year. This is another breed that is good for the beginner as they are very tame.
- Plymouth Rock – A good beginners chicken that will produce around 200 eggs a year, or one every two days, roughly. Eggs are small to medium sized and are light brown in color. These are friendly birds that are easy to tame, but because of their size, they prefer to be free range.
- Ancona – A small chicken that originated in Italy, similar in appearance to the Plymouth Rock but about half the size. It is a very skittish breed and must have its feathers clipped as it is well known as an escape artist. You can expect about 200 small, white eggs every year.
- Barnevelder – Originating from Holland this black chicken will lay about 200 small to medium sized eggs every year. It isn't a great flyer and will be happy in a run in a garden.
- Hamburg – A very pretty breed that lays about 200 white, small to medium sized eggs every year. This breed is not happy in confinement, preferring to be free range. They can get aggressive when cooped up.
- Marans – Producing about 200 medium sized, dark brown eggs

every year this is a good, docile breed. However, they aren't very tame and do not make good pet birds.
- Buff Orpington – Originating in the UK this breed will lay around 180 eggs a year, though get broody in the summer which impacts egg production. This breed is incredibly tame and makes a great garden pet. Children will love this breed as it is very social and will eat from your hand!

Dual Purpose Chickens

Sometimes you want the best of both worlds. A chicken that is a good layer but also tastes great when served up for dinner. These are some of the best dual purpose chickens you can buy.

- Leghorn – An active forager, this is a good layer, producing around 280 eggs a year. The male will weigh in at about 6lbs, with the female being a little lighter. The chickens will be ready for the table in 16 to 21 weeks.
- Egyptian Fayoumi – Well suited for hotter climates, these are disease resistant birds that are not known to be aggressive. Males weigh in at about 5lbs and females are a little lighter. They will be ready for the table in 14 to 18 weeks, though will produce about 150 eggs a year.

Picture © Joe Mabel

- Turken – A very hardy breed, this chicken is suited to hot or cold climates. They are very adaptable, active layers and good mothers. The male will weigh in at around 6lbs, and the slightly lighter

female will produce about 100 eggs a year. These chickens are ready for the table in anywhere from 11 to 18 weeks.
- Buckeye – A very adaptable breed that has good disease resistance and are reasonably docile, being well suited for colder climates. They will produce about 200 eggs a year and are ready for the table in 16 to 21 weeks. The male is about 9lbs in weight with the female being about a third lighter.
- Chantecler – Another good cold climate chicken, this is a docile breed that matures very early. They enjoy foraging and are good as a free range chicken. The male will grow to about 9lbs while the female, which is a couple of pounds lighter, will produce around 200 eggs every year. This breed is a fast grower, being ready for the table in 11 to 16 weeks.

Fattening Your Meat Chickens

You need to feed your chickens according to their dietary needs, and they will naturally put on weight. Some people will confine meat chickens and feed them a high protein diet to fatten them up, but a lot of backyard farmers do not agree with this approach, preferring to give them meat birds a happy life before slaughtering them.

You do need to be careful with broilers (meat birds) because they tend to either die of a heart attack or their legs give way due to their size! You can eat chickens that have finished laying, but you should be aware these will not be as tender as younger birds and are better suited to be cooked in a pot rather than in the oven. Older birds are good for slow cooking because it tenderizes the meat.

To fatten the chickens properly, they should have access to a good quality food for 12 hours each day and then no food for the following 12 hours. Place the water away from the food so they must move, which creates lean muscle. If you can give your chickens lots of space and allow them to forage and eat treats, then they will develop the tasty lean muscle rather than unwanted fat.

A grower food or a game bird mix will help your chickens put on weight rapidly and healthily. The typical layer feed is lower in protein and so will not encourage the weight gain you need in a broiler.

If you prefer, you can move your meat chickens to a corn based diet anywhere from three to seven days before slaughtering. Many people claim this gives the meat a much better flavor. Crack corn is best as it removes the risk of choking. A week of this diet will change the flavor of the meat and help the chicken put on weight. A constant diet of corn will encourage your chicken to get fat which will affect the taste and quality of the meat. In this last week, you can also make food available to the one(s) you plan to slaughter twenty-four hours a day to encourage last minute weight gain.

Regular treats of scratch, particularly in the last couple of weeks, will help to encourage activity and weight gain. Mealworms and other high protein chicken friendly meats will also help with healthy weight gain.

Feeding and Watering Your Chickens

Chickens are fairly versatile animals when it comes to eating, but you should remember they are not vegetarian. They will eat corn, wheat, greens, kitchen scraps, insects, worms, slugs and more. There is no right or wrong answer when it comes to feeding your chickens as everyone has their own preference. When starting out, it is best to start with a store bought, formulated chicken feed and provide your chicken with the extras they need. As you gain experience, you may reduce the amount of store bought food and feed your chickens other food. The store bought food is specially made for chickens and will contain all the nutrients and vitamins they need, so is recommended for healthy chickens.

The main feed you will give them is a pelleted food for layers, which is not too high in protein. This encourages healthy chickens that lay good eggs, without them getting fat. Meat chickens will get a grower's pellet, which is higher in protein to encourage weight gain.

These specially formulated mixes are relatively new, having first appeared in the 1970s. Some people claim it is part of a mass marketing conspiracy, but most people recognize it is due to the changes in how we live and how it is harder and more expensive to get hold of the traditional grains fed to chickens. Before the 1970s many chickens were free range, eating kitchen scraps and bugs. They were kept in rural areas, usually by farmers and homesteaders, rarely seen in an urban backyard.

However, with increasing urbanization, this just isn't possible anymore, so the insect component of the diet is minimal for most chickens, and many people do not have fresh kitchen scraps suitable for chickens these days due to changes in eating habits.

The store bought chicken feeds are designed to ensure healthy chickens that do not become overweight. For almost all of us, this is the best way to feed your chickens with insects and kitchen scraps being a treat for your chickens. Be aware that in many countries it is illegal to feed chickens kitchen scraps due to health and contamination risks.

Layer Feeds
Chickens need protein to produce both eggs and feathers. They also need high levels of calcium for healthy eggs with strong shells. The amount of protein your chickens get is very important for their continued good health. Growers pellets are higher in protein and are designed for the first few weeks of a chicken's life to help them put on weight.

When chickens molt and lose their feathers, the need for protein become more important to reduce stress and get the chickens laying again. Cat kibble, scattered in their cage a couple of times a day, is ideal as it is high in the protein chickens need during this time. Avoid using dried dog food as the protein in this is usually from cereals and is not suitable for chickens.

Commercial chicken feeds are very well researched, containing the ideal balance of nutrients, vitamins, and minerals. The layer's pellets contain about 16% protein, compared to 10% protein in wheat, which also lacks the many vitamins chickens need.

These pellets need to be available to the chickens whenever they want them, but they must be kept dry. A feeder with a rain "hat" will keep the food dry while allowing the chickens easy access to the food. Food is often kept under a shelter to prevent the wind blowing rain into the feeder.

This is the recommended food for backyard chickens because it will keep them healthy and is easiest for you to feed them. It isn't expensive, particularly when bought in bulk and is easily available. This should be the major component in your chicken's diet.

Mixed Corn

This is typically 80-90% wheat with the rest being maize and other grains. It is not used as the main food in a chicken's diet. It is considered a treat and is referred to as a scratch feed. Throw a handful of this down in the afternoon, and the chickens will go crazy scratching around and look for these treats. It gives them good exercise, keeps them interested and they enjoy the treat just like you may enjoy the occasional sweet treat!

The maize (yellow in scratch) is fattening and useful to protect your chickens from the cold. Be aware though that corn is very fattening and fat chickens are not good egg layers!

Kitchen Scraps

This is not recommended and in some countries, including the UK, not permitted. You are not controlling the food your chicken gets, and it can be a bit hit and miss as to whether they get a good diet or not. If you are feeding your chicken kitchen scraps, then it should be no more than a quarter of their diet, but be aware it is banned in many countries due to health concerns.

Greens

Chickens love greens such as grass, cabbage or cauliflower leaves, kale and other green vegetables such as spinach. Lettuce should be avoided because it has virtually no nutritional value. Avocado should also be avoided as it is toxic to chickens.

If you have a vegetable plot, growing some cabbages is a great idea. These will overwinter quite happily, and you can pick them as and when your chickens want some. Best of all, if they get caterpillars on, your chickens will love them even more!

Mealworm Treats
Your chickens will go wild for these, and they can be a great aid to taming your birds. Don't feed them too much as these are very high in protein. These are available in many pet stores, garden stores or even fishing suppliers.

Best Breeds for Meat

With concerns about hormones present in supermarket meat, more people are keeping chickens for their meat. Chickens that are raised for meat are commonly called broilers and will usually be very poor layers. Their energy goes into plumping up rather than producing eggs. Typically, these are fast growing chickens and in just five weeks can weigh as much as 4 or 5 pounds. Often by ten weeks old, they are 10 pounds in weight, which is a good size for most families. By 14 to 18 weeks you can have a good sized chicken suitable for any Thanksgiving or Christmas meal!

Male chickens are more popular for meat because they grow larger than the females. Be aware that male chickens are more aggressive than females and that they can fight, so you may need to keep them separate or keep a close eye on them. Some of the more aggressive birds will fight to the death, so you really need to be careful with cockerels.

Slaughtering and preparing your chicken can be done at home, more here, or you can send it to a butcher. The butcher will charge you a fee but will prepare the chicken, dispose of unwanted bits, and present it to you looking like it came out of the supermarket. You can ask the butcher to reserve parts you use, e.g. the liver, or discard them if you do not use them. It is best that any meat chickens are not treated as pets as it is then harder to dispatch the chicken, and children will find it particularly upsetting to see their beloved pet served for dinner.

These are some of the best breeds for meat and will all grow rapidly.

- Cornish Cross – A very fast growing breed, this chicken can reach 12lbs in six to eight weeks! They are very popular with commercial producers as well as for backyard farmers. These are not very active chickens, so they may need some encouragement, but the meat

does taste very good. There is a lot of white meat, with growth concentrating on the breast, thighs, and legs. The skin is a deep yellow color, and they are well suited for the dinner table.
- Jersey Giant – Developed to replace the turkey, this is a purebred chicken. This breed does grow relatively slowly when compared to other broilers but will reach a good 11 to 13 pounds. They are voracious eaters, which makes them expensive to keep. They are not popular commercially because of their slow growth and appetite, but they are a nice docile breed popular for home farmers. The Jersey Giant lays very large brown eggs and will be ready for the table in 16 to 21 weeks.
- Bresse – This is a popular breed, though are expensive to buy. These are white in color and well worth buying. Because the chicks are expensive, you can easily make your money back breeding the birds and selling baby chickens on. The Bresse is considered one of the best tasting chickens on the market with very tender meat. They are not the biggest meat chickens, with the male growing to 7lbs in weight on average. This is a docile breed and will take between 16 and 21 weeks to be ready for the table.
- Orpington – Although this is a good egg laying chicken, it is also popular as a meat chicken because the meat is very tender with an amazing flavor. A fully-grown female will weigh somewhere in the region of 7 to 8 pounds. Growth is relatively slow, with the chicken being ready for the table in 18 to 24 weeks. Males will typically be a couple of pounds heavier than the female bird.
- Freedom Rangers – This breed is a good forager and ideal for a free range environment, being bred to be a pasture fed chicken. They are ready for the table in as little as 9 to 11 weeks and are considered to have a great taste. Males will weigh around 6lbs with females being slightly lighter.

Food Supplements

Food supplements for humans is a massive market, and in recent years it has become trendy to get your chicken's food supplements too, but do they really need them?

During the 1950s and 60s, research was done on the nutritional needs of commercial chicken flocks with the aim to increase productivity. We now understand the nutritional requirements of chickens very well and that chickens, like humans, need a variety of vitamins to be healthy and lay well.

If you are using a modern pelleted feed, then this will have been formulated to contain all the vitamins your chickens need. Top it up with some fresh greens, scratch and the occasional insect and your chickens will have everything they need in their diet.

However, there are certain times when chickens can benefit from vitamin supplements to help them remain healthy and strong, such as:

- When stressed, i.e. moving them, changing their environment, introducing new chickens and so on
- During cold weather, when they will also require extra protein
- When breeding, though you may want a special breeding feed which does contain the required extra vitamins
- Chicks need extra help which can come from multi vitamins put in their water to help keep them strong and healthy
- Sickness – ill birds will need multivitamins to help them get through their illness and recover

There are a lot of different vitamin supplements on the market with a big variation in cost and quality. In most cases, you do not need any vitamin supplements, but if your chickens look unwell or stressed then, it can help them.

You can buy multi vitamins from garden stores, farm supply shops or online. The stock in physical stores will be much more limited, and usually more expensive than that found online. The vitamins are most commonly put in their drinking water, though some can be scattered over their food.

Water for Your Chickens

Chicken need water, particularly as they are likely to be on a dry, pelleted diet. This is absolutely vital for their health and needs to be kept clean and fresh to avoid introducing disease into your flock.

Bacteria will quickly build up in the water, particularly if the chickens walk or poop in it. It should never be allowed to turn green, and you should be careful when using disinfectants to clean the chicken's water as it could affect them.

Your chickens will need fresh water every 2 or 3 days. It should be kept in a hanging container so that they cannot stand in the water, perch on the water container or poop in the water.

During the winter months, you need to make sure that the water remains ice free, which can mean multiple daily visits to break the ice, depending on how cold it gets where you live.

In the hot, summer months the water should be kept in a shady place so that it doesn't heat up. Chickens are unable to sweat, losing heat by panting and drinking cool water. The direct sunlight can, as well as heating the water, provide a great environment for potentially harmful bacteria to grow.

Plastic water containers are available, which are more easily damaged by inquisitive chickens pecking at it. Galvanised metal containers will last longer but are more expensive to buy. The metal containers are less likely to be damaged if the water freezes but are not suitable if you give your chickens apple cider vinegar. You may also find it harder to see how much water is left in a metal container. Plastic containers, by comparison, will only last a couple of years as they break and will get damaged.

Which you buy is up to you and will depend on your personal preference.

Grit

Grit, as you know, is essential for healthy chickens and comes in two types.

- Flint/Insoluble Grit – used to grind down food
- Oyster Shell Grit – a food supplement providing calcium for healthy eggs

You must give your chickens grit because they use this to break down food. Chickens do not have teeth and to grind down their food they store grit in an organ called a gizzard. Food passes through this and is ground down to make it suitable for digesting. Free range chickens can get enough of this in the wild, but in runs and coops, they will require you to give them flint grit.

It's not expensive to buy and is easily found in pet and farm shops. Buy a good quality container to put it in so that it lasts and isn't damaged by your chickens.

Oyster shell grit is a good supplement to give your chickens. As the shells are high in calcium, it helps the chickens get enough of this mineral to produce strong shelled, healthy eggs. Most formulated feeds will contain plenty of calcium, but it doesn't hurt to provide this supplement. If your chickens produce eggs that are irregularly shaped or have weak shells, then you should immediately provide them with oyster shell grit. As this is very affordable, it is worth providing for your chickens all the time.

There are a wide variety of grit hoppers available, but you need to buy one that isn't going to get tipped over by your chickens or filled up with water. Galvanized metal hoppers are best because they last longer, but they are more expensive to buy.

Plants and Treats for Your Chickens

Your chickens will enjoy treats and fresh plants to eat, but you need to be careful you don't give them anything that is poisonous to them. Treats are especially useful when you are taming your chickens, but do not get them so used to treats that they refuse to eat their regular food.

A balanced diet is important for healthy chickens, as it is for humans, but treats are good and provide some excitement for the chickens. You should avoid giving chicks anything other than their normal food until they are two or three weeks old.

Treats should be given to your chickens as an occasional thing. A handful of scratch every afternoon is a great treat, but too much is bad for your chickens. They will prefer to eat the treats rather than their normal food if they think they can get away with it, just like children who will prefer to eat candy rather than healthy food.

Scratch is a popular treat for chickens, but there are a lot of other treats you can use as well. Some popular treats include non-citrus fruits such as apples and pears, berries, cauliflower, broccoli, cabbage, carrots, flowers, dandelions, pumpkins, and tomatoes.

Fruit is particularly good for chickens, as it is for humans. Planting fruit trees in your chicken runs (a dwarf variety is best) allows your hens to eat fallen fruit. Chickens like fruits such as pears, apples, figs, persimmon, and peaches. Don't worry if they are full of worms as the chickens love the worms too!

Citrus fruits are best avoided as they contain a chemical called limonene which is toxic to chickens. Although it affects their health and ability to lay it is unlikely to kill them, but it is best avoided just in case.

Chickens really like strawberries and other berries too. If you allow your chickens to roam in your garden, then keep your strawberries well out of their reach because they will strip the plants of fruit! Berries are very good for your chickens because they are high in vitamins. Elder trees are worth planting by runs because they cast shade and provide a delicious berry for your chickens.

Chickens can also eat rose hips, sloes, and hawthorn berries, which are easily found in September and October when other berries are hard to find. Often you can find them growing in hedgerows and collect a few handfuls for free!

Cruciferous vegetables such as cauliflower, broccoli, and cabbage are known as a superfood for humans because they are jam packed with vitamins. They are also very good for your chickens! Some chickens will go crazy for these whereas others will be indifferent. Until you try your chickens on them, you won't know which way yours go.

These dark green, leafy vegetables are very high in calcium (6.3% in cabbage) which is great if your chickens produce eggs. A good treat for chickens is broccoli florets mixed in with some sliced cucumber and grated carrot ... they'll go crazy for it! Even grated carrots by themselves make for a good vitamin packed treat for your chickens. Carrots are also very easy to grow at home, providing a cheap source of treats for your chickens.

Dandelions are considered a weed, though they are eaten in some parts of the world. Chickens love the leaves, so you can grow them in your run or elsewhere in your garden or collect them from the hedgerow or other people's gardens. Just be careful there are no pesticides or weed killer on them as that is toxic to chickens.

Chickens find flowers very tasty, and if you grow them in pots they will dig out the flowers to use the pot as a dust bath! They like flowers such as marigolds, nasturtium, rose petals, chrysanthemum and more. You can easily grow these at home, or you can buy flowers from stores after checking they haven't been treated with insecticide. Nasturtium is very easy to grow and produces a gorgeous trailing mass of color ... you can eat the flowers, leaves and the berries, which are sold commercially as capers.

Lettuce is a good treat for your chickens, but you shouldn't give them too much as it is empty calories. Suspending a lettuce head in your chicken run gives them something to do as well as a treat. They'll enjoy pecking at it and tearing it to shreds!

Both melon and watermelon are a great treat for your chickens and in the summer months helps to keep them hydrated. They will dig away at the melon, scooping out both seeds and flesh. Pumpkins are likewise a great treat for your chickens, with pumpkin seeds being very high in protein. You can even give your chickens your carved pumpkins after Hallowe'en, and they'll enjoy tearing it apart and eating the flesh.

Chickens love tomatoes but keep them away from the plants as they contain tomatine which is dangerous to chickens. The tomatoes themselves are high in lycopene which is an antioxidant that has scientists very excited about its use in treating humans for a wide variety of complaints. Too many tomatoes can upset their stomachs, but in moderation, they will enjoy a handful of fresh tomatoes every now and again.

Your chickens are probably smarter than you give them credit for, no matter how dumb they act. They will know when something is poisonous to them and will typically avoid it. As they are natural foragers, they have a lot of different plants they can eat, but there are a few plants you should avoid.

What to Never Feed Your Chickens

The following plants are some of the common plants known to be poisonous to chickens. You should ensure none of these grow in or near your chicken run and you should never feed these to your chickens.

- Bloodroot
- Bull Nettle
- Bracken
- Bryony
- Carelessweed
- Castor Bean
- Cocklebur
- Curly Dock
- Delphinium
- Fern
- Foxglove
- Ground Ivy
- Hemlock
- Horse Chestnut
- Horse Radish
- Hyacinth
- Hydrangea
- Ivy
- Laburnum seeds
- Lantana
- Lily of the Valley
- Nightshade (Also called Deadly Nightshade)
- Rhododendron
- St. Johns Wort
- Tulip
- Water Hemlock
- Yew

Blue green algae, which is often found in ponds or neglected water containers is deadly to chickens. Ponds in free range fields and runs, as well as water containers should be kept clean and free of this dangerous algae. Regular cleaning of containers and making an effort to keep ponds clean will help your chickens stay healthy.

Bulbs such as daffodil and tulip should not be eaten by chickens. It is thought these are poisonous, but at the very least they will make your chickens ill. Chickens will attempt to dig these up, so be careful planting bulbs anywhere your chickens roam.

Your chickens are quite adaptable creatures that naturally have a varied diet. Feed them the occasional treat to keep them healthy and interested, and keep them away from things that are bad for them. With a balanced diet and the odd treat, you should find chicken keeping to be very trouble free.

Managing Your Chickens

Chickens are generally relatively low maintenance creatures. They will need visiting twice a day; once to let them out of their coop in the morning and once in the evening to put them away again. At those times, you can do most of the jobs relating to them and then once every 7 to 10 days you clean them out fully. Apart from this, you can let the chickens lead their own lives. The main portion of work is before you get the chickens, building their living quarters and run.

This section will tell you about many of the tasks involved in keeping chickens to help you understand fully what is involved and what needs doing.

Cannibalism and Pecking

When we discussed breeds earlier on in this book, you learned that some breeds are more aggressive than others. Even within a breed, you have a pecking order, which is the social hierarchy amongst the chickens. There will be one or two chickens that feel they are the leaders of the flock and others which are more submissive. This will often change over time and usually involves fighting.

Chickens will naturally peck at each other to establish the social hierarchy, hence the term pecking order. Usually, this just involves pecking and pulling out feathers, which results in no more damage than a small injury to the skin which usually heals without any trouble.

In some cases, this can lead to cannibalism. During research, it appears that laying hens are more like to peck at short feathers rather than long ones. It also appears that hens which lay brown colored eggs are much more likely to engage in feather pecking than those that lay eggs with white shells.

Feather pecking occurs whether your chickens are free range, kept in a coop or allowed access to a run. This activity is more common when chickens feel they are confined and overcrowded. In smaller flocks, there are fewer pecking problems.

Cannibalism occurs when the skin is torn and eaten by other chickens and can happen in any living condition and any breed. Unfortunately, this is a learned behavior and it will spread through the flock and can result in the loss of a lot of your birds. In worst cases, you have to euthanize your entire flock because they have developed a taste for chicken flesh.

This is easier to prevent than to cure, so keeping your chickens happy and healthy is more likely to stop cannibalism from happening.

Overcrowding

One of the main causes of cannibalism and feather pecking is overcrowding. Although you may have a large run and feel you can keep lots of chickens when the weather is bad, or they are shut in their coop at night they could end up feeling cramped.

It is vital you provide more than enough space for the birds to roost, nest, eat and drink. If you do not, then you will encourage competition, which will lead to fighting for position in the flock as the more dominant birds protect what they consider to be a scarce resource. Make sure there is plenty of room for all your birds to eat at once, which will stop the more submissive birds from losing weight and being attacked by other chickens.

Picture © Winner man

The coop needs to have enough floor space for all your chickens to be inside at once and plenty of space for birds of all social classes to roost.

Mixed Breed Flock

Some breeds of chicken are naturally more aggressive than others. If you want to have a variety of breeds in your flock, then you should get ones that are of a similar temperament and watch them closely.

If you choose a more aggressive breed and a more docile breed, then the dominant birds will bully the more placid birds. Keep the different breeds separate which will stop any conflict.

Overheating

When chickens get too hot, it makes them uncomfortable, and they are more likely to start pecking each other. Providing good ventilation in the hen house and adequate fresh water will help them stay cool. During hot weather, you may want to change their water at both morning and evening visits.

Too Much Light

If you use lights with your chickens, then you need to be careful as if the light is too bright or they have too much light then the birds will get aggressive towards each other. If you are brooding your hens, you should not use a white light bulb greater than 40W. Should more powerful bulbs be required, then red or infrared bulbs must be used.

Your hens should not have more than 16 hours of light each day as constant light is very stressful. Broiler birds are often given 16 hours of light (1/2 to 2-foot candle intensity) followed by eight hours darkness.

Poor Nutrition
Your flock needs good quality food and plenty of water. If your birds have a diet that is too low in sodium, phosphorus, and protein, then they will tend cannibalism. A high energy, low fiber diet will cause your chickens to be more active as well as more aggressive.

This is why it is important that you feed your chickens a healthy, balanced diet. The pelleted, complete foods contain everything your chickens need and will keep them happy. Remember that as chicks grow into adults so their dietary requirements change and you need to change to the appropriate food for them.

Chickens are natural foragers. When they are unable to do so, they can get frustrated and direct their foraging behavior aggressively towards other members of the flock. Provide your chickens with an environment and materials that can mimic their foraging behavior such as grass clippings or straw that will encourage them to scratch around. Throwing scratch down every afternoon can also help the chickens behave better as it gives them something to do.

Chickens preen themselves regularly to keep clean, using a preen gland near the tail. This gland contains oil which is salty. If your birds have a diet that is too low in salt, they will overuse this gland, damaging their feather and even peck at glands on other birds.

Injured or Dead Birds
Chickens are naturally attracted to blood. If a bird is injured, then cannibalism can break out as other birds peck at the injury. You must ensure that there is nothing in their coop or run that can damage the birds. Check the birds carefully, and if any is injured, it should be isolated from the flock until it is healthy again. If a bird dies, it needs removing immediately to prevent cannibalism.

Intermediate Flock Size
The social hierarchy within a flock depends on recognition of individual hens. In a large flock, the birds are unable to recognize all the other chickens in the flock, so the social order breaks down as they become more tolerant of other. This happens at a flock size of about 30 chickens.

In smaller groups, the chickens can recognize each other and so they are all aware of where they stand in the pecking order.

In flocks that are too big for a stable social hierarchy to develop, but too small for the tolerance to kick in you end up with social problems.

Therefore, you either need a flock greater than 30 birds or much smaller. Most backyard farmers will have less than ten chickens, often around 4 or 5, so this isn't going to be a problem.

Age and Color of the Flock

Having a flock where the chickens are a wide variety of ages, colors, sizes and breeds will upset the social order, unless they have been raised from chicks together.

Avoid mixing birds like this and avoid mixing birds with different traits. Birds that are bearded, crested or feather legged can make birds without these traits curious, which will lead to pecking.

Abrupt Changes

You can cause your chickens stress through moving them, changing their environment or changing your management practices. Help your chickens adjust to changes, and it will reduce the risk of the developing cannibalistic behavior. If you are changing the feeders and water containers, then leave the old ones in place for a few days, which will help them to adjust. Remember that chickens do not like change so try to minimize any stress they may experience.

Poor Nest Boxes

Your birds need enough nest boxes and good laying conditions to prevent them fighting and engaging in cannibalism. Avoid placing bright lights near the nest boxes; your chickens need a safe, dark place to lay their eggs.

Picture © Andrew Bone

Assessing and Monitoring

You should keep an eye on your chickens to monitor any feather pecking. Whether you take written notes or keep a mental note will depend on your memory and the size of your flock.

Check your chickens and look at the quantity of damage and the severity of it, if there is any at all. If you see a chicken getting worse, then it must be removed it from the flock and isolate it until it has recovered and you can reintroduce it. Watch it carefully when you do put it back in the flock and make sure it doesn't get pecked again.

Some breeds are more prone to cannibalism than others, so choosing the right breed is very important.

If cannibalism breaks out in your flock, take immediate action to prevent it and stop it spreading through the flock. You can remove the aggressive birds and isolate them. The victims of cannibalism should also be moved and cared for in isolation until well enough to be reintroduced to the flock or humanely euthanized.

Having an environment where the chickens are enriched through activities, particularly relating to foraging will help to minimize this problem. Make sure there is sufficient room to perch and nest, plus plenty of food and water to help prevent competition for resources and fighting.

In the worse cases, you can trim the beak, which is banned in a number of countries as it is considered a cruel mutilation, or you can fix goggles to the beaks of the more aggressive birds, which is the more humane solution.

Hanging a lettuce, head of broccoli or cabbage from a string in the run makes for an effective enrichment device to divert pecking away from other chickens. Move these regularly to help keep them interested.

If your chickens are bored, then they are more likely to fight. Keeping them occupied and entertained will ensure your birds are healthy, happy and content.

Protection from Predators

Protection from predators should always be foremost in your mind. You can easily lose your entire flock overnight to a hungry fox. When you build your chicken coop and run you must take steps to predator-proof it. However, you need to keep a close eye on your security measures as over time things can wear out, break or be worn down by predator attacks.

Firstly, your chickens should always be put in the coop at night, which is closed securely behind them. If you raise your chicks in the coop, they will naturally return to roost at night. Make sure the coop is about a foot off the ground which will discourage pests and predators living underneath it. You should also regularly check the coop for damage and make good any necessary repairs.

Electric fences may be the best solution in areas where predators are common. Remember that many of the larger predators are either good diggers or good jumpers so you must prevent them from gaining access to your chickens. Building six-foot fences around your run with a roof on will prevent predators jumping into the run.

If raccoons are a problem in your area, you will need a tight mesh wire as they can reach through wider meshes and kill your chickens.

A motion sensor activated night light will help to scare away predators, though make sure it doesn't disturb your chickens. If you have a cover free area perimeter around your chicken run this will also help discourage predators as they dislike being exposed. Plant bushes inside the run, but keep the outside clear so predators cannot hide and annoy your chickens.

Worming Your Chickens

Chickens suffer from intestinal worms which is most noticeable if they struggle to gain weight or suddenly lose weight. These worms can cause health problems, and if your chickens are kept in one area, they will be infecting each other as they shed worm eggs in their feces.

One of the most effective wormers on the market is Flubenvet, which your chickens should take every 3 to 6 months. This is a proven wormer that is licensed for use in chicken feed. If you have a worm problem, this is one of the best ways to kill of these infectious parasites.

Vermm-X is approved as a wormer in organic system and is an herbal product, which your hens will need every month. Between this and Flubenvet plus good hygiene, you should be able to keep your hen's worm free.

Try rotating the chicken's grazing area every week or so as this gives the ground a rest and prevents the build-up of worm eggs. Of course, this may not be possible for everyone to do. Keeping the run and hen house clean and free of feces will go a long way to prevent the spread of worms.

If you have a particularly bad case of worms, then your chickens will need two doses of Flubenvet within a three week period to clear up the worms and prevent immediate re-infection.

This popular wormer comes as a powder which is mixed into the chicken feed, using the instructions that come with the product. You can buy pre-mixed feed which is a lot easier to use and can work out cheaper!

Wing Clipping

Most chickens are not particularly good at flying. They don't tend to fly as a normal bird does, but can certainly fly well enough to get out of a run or up into a tree. If you do not have a roof on your run or your chickens are free range, then clipping their wings is the best option to stop them escaping and get into danger and making a pest of themselves.

Keeping Chickens for Beginners

Picture © ErikBeyersdorf

You do not have to clip both wings; just one is enough to stop the bird from flying. It is a painless process that is quite easy to do when you get the hang of it. Clipping a single wing unbalances the bird as one wing is shorter than the other and stops them from flying. Initially, you will be nervous doing this, though if your chickens do not go outside of a roofed cage, you do not need to clip their wings unless they are flying into the cage and hurting themselves. Some breeds try to fly more than others, so your chickens may never actually try to fly.

The hardest part of the process is catching the bird. Use treats such as mealworms or grapes to attract the bird to you. Then you hold it with the chicken facing towards you with its chest resting on your forearm and its feet between your fingers. This gives you a free hand to hold the chicken and clip the wing. For your first few times, you may want a handy helper to hold the bird so you can concentrate on cutting the feathers.

Picture © Mark Robinson

Once the bird has settled, you fan out one of its wings ready to trim the first ten primary flight feathers (the longest feathers). The other feathers are not touched so that it can still stay warm.

The feathers must be cut to the right length; cut them too short and you hurt the bird, and if they are too long the bird will still be able to fly.

Look closely at the feather, and you will see that the feather has a blood supply to it. Underneath the wing, you will see dark shading on the feather shaft where the blood is. Avoid cutting this dark area as it hurts the bird and encourages infection.

You are safe to cut where the shaft is white. Use sharp, clean secateurs or scissors and cut along all ten of the primary feathers. Aim to cut just below the smaller feathers that overlap the primaries, without damaging the smaller ones. You should take off no more than about 6cm. There is no point trimming both wings, so make sure you just do one.

The chicken is going to be flustered by this undignified treatment, so give her a little while to settle down before letting her rejoin the flock. Her agitation will spread to the rest of the flock so you may have to give them a few minutes to calm down before clipping the next bird. Releasing her straight back into the flock will get them all squawking, and the next one will be hard to catch. If you've got a lot of chickens or struggle to tell them apart, you may want to mark or separate them until they are all done.

You should do this around every six months or so, depending on how quickly the feathers grow back. In all fairness, when the chickens realize they are being fed and looked after, all but a few breeds lose the will to escape as they enjoy being taken care of!

Exercise Requirements

Chickens need exercise to stay healthy, both physically and mentally. If they cannot move around, they will get ill and aggressive. This is why a run is so important for chickens. You may not be able to let your chickens roam free range, but you can certainly build a run so your chickens can get the exercise they require.

For a run, you will need a minimum of a square meter (three square feet) of space per hen. If you can give them twice this room, then they will be much happier. The key is not to overstock your run. For most backyard gardeners, somewhere between three and six hens will be more than enough. Therefore, for six hens you will need six square meters of floor space, with twelve square meters being better. The larger area gives you more flexibility and ensures your hens are happy and productive.

A double decker hen house, raised up off the ground, is a good space saver because the space under the coop can be a sheltered outdoor area. It means the hens can still go outside when the weather is bad, and they have somewhere to hide from it.

Most people will not have the space to move the run to fresh areas of ground regularly so you can expect the soil to get muddy, particularly in damp weather. Covering the ground with several inches of scratching material, e.g. hardwood chips, is easy to replace and provides a safe flooring during inclement weather. Sand or gravel are other options, though make sure your chickens have regularly fresh greens.

If you are short on space them look at the bantam breeds, which are much smaller and do not need as much room. They will still produce eggs but don't need as much space as the big breeds like Leghorns.

As part of their exercise regime, they will need to be entertained to prevent anti-social behavior such as feather pecking. Chickens, though, are not difficult to amuse!

A dust bath is important for your chickens, though many people do not provide one. A plastic box with some play sand (not sharp sand) will make for a great dust bath and help keep your chickens healthy.

Fresh green food is important for healthy chickens and having to forage, jump and hunt out these tasty treats is a good source of exercise.

Put some logs or branches in their run so that they can perch and climb during the day. Again, this is good exercise, and the chickens will enjoy exploring and playing on them.

Introducing New Chickens to the Flock

Every flock has a social hierarchy or pecking order that the flock accepts and lives with. When you introduce a new hen or hens to the flock, this upsets the pecking order as none of the hens know where the new chicken sits in their social hierarchy. You end up with the chickens jostling for position and fighting.

You may want to expand your flock or replace a chicken that has died or made its way on to the dinner table. It is, therefore, quite likely that at some point you are going to want to add new chickens to your flock.

One method used by breeders is to put the new chicken behind a fence or small run near the original flock. This gives them the chance to get used to their new environment and the other chickens.

Introducing the new chicken into the coop at night will prevent fighting initially, as they will not fight at night. It gives the chickens a night to get used to each other.

Pecking is likely going to take place, so just leave them to get on with it and establish the pecking order, but keep a close eye on all your chickens. If any blood is drawn, immediately remove the injured bird as chickens are attracted to red things and will continue to peck. Once the hen is healed, try putting her back in the run again, but keep an eye on them all.

Often it will be one particular hen that is causing the trouble. Try removing the trouble maker until the new hen has settled in the flock, then put the trouble maker back with the others and keep an eye on it.

As a last resort, you can use an anti-peck spray or put a bumper on the beak of the bird. This doesn't interfere with eating or drinking but stops the end of the beak closing completely which prevent pecking and damaging other chickens.

Routine Jobs When You Own Chickens

There are quite a few routine jobs to perform when you own chickens. Although they aren't high maintenance, they do need regular attention to keep them healthy. Here are some of the common routine tasks and the frequency you need to do them.

Daily Tasks
These must be done every day.

- Provide fresh food and water, topping up the pelleted food feeder as required, checking the food hasn't gone moldy. Discard any food that is moldy.
- Clean water containers before refilling to prevent a build-up of potentially dangerous algae.
- Spend a minute or two checking your hens and their environment to make sure nothing is wrong, looking out for blood or feathers on the floor or damage to any fencing.
- Collect eggs and watch out for broody hens.
- Scatter a handful of scratch in the afternoon as a treat.
- Close the coop entrance every night, opening it in the morning, to protect the hens from predators. The coop should be opened as soon after dawn as possible and closed at dusk when the chickens go to roost.

Weekly Tasks
The following tasks should be performed every week.

- Clean out the coop and run to prevent the build-up of waste matter which causes disease and respiratory problems.
- Between May and October check cracks and the ends of perches for red mite.
- Thoroughly clean feed hoppers and water containers.
- Provide fresh greens a minimum of twice a week if the chickens are not free range.
- Check and top up grit hoppers.
- Check a couple of birds at random to make sure they are healthy, particularly around the vent area where lice lurk.

Monthly Tasks
These jobs are done once a month.

- Provide the chickens with 2% Apple Cider Vinegar in their water for one week every month. This acts as an antibiotic, reduces stress and has vital vitamins and minerals in.
- Provide the monthly dose of Verm-X to keep your birds worm free.

Month by Month Task List

As well as the regular jobs, you need to do; there are other jobs that you can get on with, depending on the time of the year and what you do with your chickens. The following list tells you what needs doing every month, though if you don't breed chickens some of these won't apply to you.

January

- Test your incubators and decide which chickens to breed. Those aged 2 or 3 years old are good breeders.
- Add a vitamin supplement to water if the weather is cold to give your chickens extra protection.
- Rub Vaseline or petroleum jelly into the combs of large combed birds to protect them from frost bite.
- Provide regular greens if the ground is frozen and the birds cannot get at grass.
- Clear snow from the front of their house so they can still get out.

February

- Order supplies for incubating chicks, including spare heat lamps.
- Continue to add vitamin supplements to water for breeding chickens or if it is cold.
- Continue putting petroleum jelly on combs.
- Provide regular greens.
- Check eggs in the incubator for fertility.

March

- Collect eggs and incubate those you wish to hatch. All your hens should be laying, and fertility should be at its best.
- Add gravel, sand or wood chippings on the ground in case it turns to mud in wet weather.
- Worm the whole flock using Flubenvet.

April

- Give the hen house a thorough spring clean.
- Tend to hatched chicks and keep an eye on broody hens.
- For those breeders with a cockerel, put poultry saddles on your female birds, so they don't lose their back feathers.
- Build new housing as required, remembering that young chicks and older birds should not be mixed.

May
- Start preventative red mite treatment.
- Let young chicks out into grassy runs and lock them up at night.

June
- Continue the preventative red mite treatment.
- Inspect fencing thoroughly as the next few months is peak fox attack time.
- Ensure there is plenty of shade for your birds as it starts to get hotter.
- Move their water into a shady area so it can be used to cool them off.

July
- Continue preventative red mite treatment.
- Treat or paint any woodwork.
- Sex your chicks and remove unwanted cockerels.

August
- Continue preventative red mite treatment.
- Give the chickens extra vitamins while they molt and a handful of cat kibble for extra protein.

September
- Continue preventative red mite treatment.
- Move this year's chicks into their permanent home or integrate them with your existing flock.
- Worm your whole flock with Flubenvet.
- Repair runs, houses, and fences before winter.

October
- Stop preventative red mite treatment as the temperature drops.
- Review your stock levels, ordering anything needed for winter.
- Young cockerels will start fighting next year should be removed.

November
- Ensure you have enough water containers to last you through the big freeze.
- Make sure you draft proof your chicken coop

December

- Add a vitamin supplement to the chicken's water on alternate weeks.
- Provide regular greens and extra corn to help the chickens stay fat and keep warm.
- Ensure your chickens can get shelter from wind, rain, and snow.

Seasonal Precautions

So long as you have chosen the right breed for your climate, you will not have any problems during cold winters or hot summers.

People often wonder about heating their coops during winter, and this isn't a good idea. Chickens are naturally able to adapt to cold weather with their metabolism changing along with the seasons. If you heat their coop, then they will not adjust to the cold outside temperatures which mean they never learn to adapt to the cold. This can cause problems if your heat stops suddenly or they get caught outside for any length of time. A simple power outage on a cold winter's night could cause your entire flock to freeze to death because they have no natural protection against the cold.

Picture © seppingsR

Any chickens that have combs need to have Vaseline or petroleum jelly put on their combs and wattles to protect them from frost bite, which is very painful to the birds and needs to be avoided.

You must ensure their water supply doesn't freeze as chickens cannot live very long without fresh water. Either bring the water indoors when you put the chickens to bed or check it a couple of times a day to make sure it hasn't frozen over.

In warmer months, hot weather is a real risk to your flock so make sure they have shade outside, plenty of fresh water to keep them cool. Also, ensure there is adequate ventilation inside their coop, so they don't get too hot. You may want to consider putting a fan in their coop to help circulate air, particularly at night.

In both hot and cold weather, you can expect egg laying to reduce. When temperatures normalize, normal laying rate will resume.

During the winter months, you can expect your chickens to eat about 10% more than they would usually to bulk themselves up. Provide them with extra food as well as cracked corn and greens, so they have plenty of food to keep them warm.

Chickens keep warm by roosting together, so you need to ensure that they have enough perch space to huddle together. If you have an odd number of chickens, ensure that at least one perch is big enough to accommodate three chickens, so none are left by themselves to get cold. Make sure you do not block the vents in winter as the air circulation is required to prevent the build-up of noxious ammonia inside the coop.

During the winter months, you can move your coop, so it is out of the wind and sheltered from the cold. This may not always be possible so you can line the walls and roof with cardboard as insulation. Just staple or tack it in place and remove it in the spring. Straw (not hay) is a good insulator for the floor.

You can provide lighting in the coop to keep the chickens laying during the colder months, though most home keepers won't go this far. Make sure the chickens get about 16 hours of light a day and that it isn't too bright. It should not shine directly on the chickens and is best bounced off the roof. You need to make sure that it does not heat the coop too much. They will then lay in the winter months, but most people don't do bother with lighting as they allow their chickens to rest and build up their strength over winter.

As chickens are not big fans of snow, so they may not want to go out much in winter. Put a roof on the run to prevent snow building up so they can still go outside and get the exercise they need. If you cannot put a roof up, then clear snow by hand, so they get out of their coop.

CLEANING THE COOP

Cleaning is a big part of owning chickens, so you may as well get used to it. Depending on the size of your flock and set-up you may have more or less cleaning to do.

There are two regular cleaning jobs that need to be done:

- Cleaning the chicken coop
- Cleaning the run

Let's discuss these in a little bit more detail, so you know exactly what is involved.

Cleaning the Chicken Coop

This is a very important job, usually done once a week, which keeps your chickens healthy. How much you need to clean it will vary according to how much time your chickens spend inside, your nose will be your guide here. In warmer months, the chickens will spend less time inside, and so the coop needs less cleaning whereas in the colder months when the chickens are indoors more it will need more frequent cleaning.

Chicken droppings give off ammonia which is harmful to their respiratory system. Generally, if the smell of ammonia is overpowering, then your hen house is well overdue a cleaning!

Cleaning the Coop

Essential tools for cleaning your coop are:

- Large bucket
- Small shovel
- A paint scraper
- A stiff brush
- Chicken friendly disinfectant
- Heavy duty rubber gloves
- Face mask (if you don't like the smell of ammonia)

Remove the perches and droppings boards from the coop and clear all the bedding out. This can be put into your bucket or directly into a garbage bag. Remember to clean out the nest boxes too. Unsoiled nest box material can be used on the floor of the coop.

The chicken bedding is very good on the compost heap, which is a good way of making use of it rather than throwing it away.

The scraper is used to remove dried droppings from anywhere in the hen house. This can be hard work, but if you do this regularly, there won't be too much to remove. If you find piles of gray dust that look like cigarette ash, then this is an indication of a red mite problem so a deeper clean, see later, is required.

When you have cleaned out the coop, you can then clean it with a disinfectant or use a poultry disinfect powder. If the weather is okay, leave the doors open for an hour or two to air out the coop. Your chickens are bound to try and help you with the cleaning, but try to keep them out of the way so you can get on with the job.

Add fresh bedding and nesting material and replace the perches and droppings board. You can spray the coop with an anti-mite powder or sprinkle Diatom powder around, rubbing it into all the nooks and crannies.

Cleaning the Run

A lot of people have a static run, which will become a haven for worms and possibly other diseases. The smaller the run, the more often it will need to be cleaned.

In dry weather, it is easy enough to rake up the droppings and compost them. Lining the run with hardwood chips or sand is recommended as it prevents the run turning into a churned mud bath, which your chickens will not appreciate in wet weather. Depending on the weather and the number of chickens, you will need to change or clean the substrate regularly. Some people use rubber chips on top of the soil because they are relatively easy to clean and do not need replacing, unlike wood chips.

In the run, you should use a ground sanitizer and a poultry disinfectant to destroy any worm eggs and prevent diseases taking hold. It's best to put your chickens in their coop while you do this otherwise they insist on helping and getting in the way.

Spring Cleaning

It's a good idea to give your run a deep clean a couple of times of years. Most people give it a big spring clean early in the year before red mite season starts and then another as winter approaches when red mite season ends. If you do suspect a parasite infection in your hen house, then you need to give the whole house a spring clean to get rid of these pests.

It is best to do your spring clean on a warm, dry day and to start early on as it can take a while. This means you can put the chickens in the run while you clean and everything has a chance to air and dry completely.

Firstly, take the coop apart as much as you can. Then remove all the bedding, empty the nest boxes and scrape off the droppings, just as you would in a normal clean.

Use a cleaning product known as Poultry Shield to clean every nook and cranny of the hen house, paying particular attention to grooves, corners and edges. This cleaner is safe to use around chickens and also destroys red mite. Mix it up as per the instructions on the container and use a stiff brush to get into all the corners. Where you cannot get to with the brush, use a sprayer to ensure you cover the entire coop with the spray.

Once this is done, leave everything to air out and dry in the warm weather. The ultraviolet light in sunlight kills a lot of bugs, so leaving the door open whilst it is drying is a good idea.

Choosing Chicken Bedding

What you put on the floor of the coop isn't technically bedding as the chicken's sleep on the perches, not the floor. If we are being accurate, it should be called floor litter as it is there for insulation and to absorb the moisture from the chicken droppings.

Picture © furtwangl

For most of us, the choice of floor material is influenced by its compostability. Many backyard farmers will also grow vegetables. The chicken manure is a great source of nutrients for plants so you need to ensure that whatever bedding you choose can be put on your compost heap and will rot down.

Whatever you choose to use it needs to be dust free as dust can damage the respiratory systems of your chickens. Dust extracted wood shavings, like those used for small mammals like hamsters and mice, work very well, particularly as they are cheap when bought in large bales. They are also easily available from pet, garden and farm supply stores.

Shredded hemp is a popular choice where it is available because it is not only very absorbent and dust free, but it also repels flies. This, though, is not available everywhere and is more expensive than wood shavings.

There are wood products that are designed for chicken bedding. These are a little bit more expensive, but is dust free and easily composts.

Another popular option is straw. Normal straw, though, is quite dusty and isn't very absorbent. You can buy dust extracted, chopped straw which makes for a much better bedding. Avoid hay at all costs (they are different), as hay contains mold spores which can be harmful to your chickens and damage the coop. Shredded bark should be avoided for the same reason.

Shredded paper is not recommended because it is a poor insulator, doesn't absorb well and blows away in the wind! When you open the coop, it will blow all over the run and then it is more work to pick it up. Shredded cardboard though works well as it is much more absorbent. You can buy a shredder and shred your waste cardboard, so long as it is clean.

Which bedding you use will depend on what is available to you. My personal favorite is wood shavings because they are so absorbent and insulating. They also break down very easily on the compost heap.

Uses for Chicken Manure

You are going to end up with a lot of chicken manure and, although you can throw it away, it is very useful to you as the backyard farmer. Poultry manure is one of the most popular non-chemical fertilizers available, used by vegetable gardeners across the world.

Chicken manure is very high in nitrogen, which is the main nutrient plants need for growing healthy, leafy growth. It also contains smaller amounts of other vital nutrients.

Picture © Herzi Pinki

Compared to synthetic fertilizers, it is low in nutrients, but it is organic, free and readily available. Pelleted chicken manure usually has around 4% nitrogen, 2% phosphorus, and 1% potassium. Chemical fertilizers are often 2 or 3 times this.

Fresh chicken manure can be applied as a top dressing for your plants that need a lot of nitrogen such as vegetables, plums, and black currants. It can also be added to your compost heap where it will break down naturally. Try to intersperse layers of chicken manure and wood shavings with other materials, so you have 2" of chicken manure then 6" of other compostable material.

Be aware though that fresh chicken manure does contain bacteria which is harmful to humans. You must always wear gloves when handling manure and ideally wear a mask to prevent you breathing in dust and ammonia. Avoid eating and drinking while working with chicken manure. If there is an outbreak of bird flu in your area, then you should not compost your chicken manure until you get the all clear.

Some people with large quantities of chickens actually sell the manure in pellet form as a way of financing their chicken habit. Dried chicken manure is extremely popular with vegetable gardeners, so it is easy to sell. If you cannot pellet it, you can always store it in black plastic garbage bags and give it away to local vegetable growers. They will be pleased to get a free fertilizer, and it means you don't have to try to dispose of it yourself.

In general, chicken manure has a pH of between 6.5 and 8.0, depending on the age of the manure, the age of the birds, the bedding material and the diet of the birds. As it tends towards the alkaline side, it isn't suitable for plants that like acidic soils such as blueberries, rhododendrons.

Be aware that fresh chicken manure will actively break down when you put it on the soil, generating heat as it does. This can damage plants of any age, though young plants are particularly vulnerable. Avoid this scorch by not planting directly into fresh chicken manure, as it will burn the delicate roots of plants. Avoid getting it on stems or leaves for the same reason.

Leaving the manure for a few weeks to break down before applying it can remove the risk of scorch damage to your plants.

Pests, Diseases & Predators

Chickens are susceptible to disease, and you need to keep a close eye on them, checking them on your daily visits. Unfortunately, it is quite hard to spot any sign of illness in a chicken as they show few signs until they are seriously ill, at which point it can be too late to treat them.

Over time you will become familiar with your chickens, and you will know when they are not acting right. Pay attention to their activity levels, the color of their combs, the state of their eyes and you will soon become familiar with their general state of health. This means that when there is a subtle change indicating an illness, you will be able to spot it. Regularly handling your chickens will help you to know their plumage, what normal skin and feet looks like and so on, which makes it easier for you to spot any differences.

An easy way to check a chicken's health is through its weight. Feel for the keel bone, which is the middle of its chest. There are muscles on either side of this so it should feel flat. If it feels pointed and there is a dip on either side of it, then your chicken is underweight. If there are bulges on either side, then your chicken is too fat and needs to diet!

Like many animals, when they are feeling unwell they either stop eating completely or eat less, which leads to weight loss. Tail dropping is another sign of serious illness in a chicken as the nervous system is the first thing to be affected by illness. If the chicken's tail is bobbing up and down as it breathes, then you have an emergency situation on your hands and should get straight to a vet.

Hopefully, your chickens will stay well, and you won't have any problems. This chapter is going to give you more information about the potential illnesses and pests that can affect chickens and help you to spot the signs of disease. You will also learn what steps to take to minimize the risk of problems and how to treat your chickens.

These are the most common problems you will encounter.

Molting
This is a yearly event for chickens and takes a lot of energy to shed and regrow the feathers. It is very tiring for the chicken, and if it happens at the same times as another stressful event, such as new chickens arriving or moving home, then your chickens will be at a higher risk of disease because their immune system is suppressed for the month or two they are molting.

To regrow feathers, chickens need high levels of protein. It will help them to provide a handful of dried cat kibble every day. This is very high in protein, and the chickens will love the treat. Do not use dried dog food as the protein comes from cereals and is not as effective.

During this time, you need to ensure they have plenty of food and it is best to add a vitamin supplement to their water. Keep their stress levels to a minimum, and they will soon return to normal.

Worms
Chickens are affected by round and tape worms which can live in the intestine, stomach and wind pipe. In the latter case, this causes respiratory problems and can lead to weight loss.

Worms, wherever they are located, will lead to weight loss over time and even anemia if the infestation is bad enough. If your chicken is anemic, then the comb and/or wattle will become paler. You may also see a decrease in the quality of eggs, with brittle shells.

Worms need worming twice a year. During the summer months, they can benefit from an organic or herbal wormer being given every month. Good cleanliness of both the run and coop will help prevent re-infection of worms.

Mites
Chickens are affected by a number of different mites that can infect different areas of the chicken.

Dermanyssus gallinae – *Red Mite*
A very commonly found mite that is a serious problem, so serious they get a whole section to themselves here!

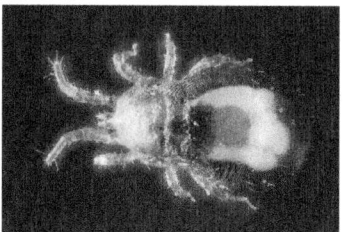
Picture © Gilles San Martin

Ornythornyssus sylviarum – *Northern Fowl Mite*
This black mite causes scaling on the face, wattles, and comb which is uncomfortable for your chicken and can cause anemia. The base of the feathers is also affected, where they grow out of the chicken, and can look like they've been chewed.

Unfortunately, these are virtually impossible to get rid of. Both the chickens and the coop need to be treated. To be honest with you, many people literally end up burning the hen house and buying a new one because these are so difficult to contain.

Cnemidocoptes mutans – *Scaly Leg Mite*
This, not surprisingly, shows up as scaly legs on your chicken. The mite burrows between the scales and causes a lot of discomfort. In worst cases, your chicken will struggle to walk. Usually, this will start off in older birds and then spread to the rest of your flock.

Using Epsom salts diluted in warm water, soak off the dead skin, careful to avoid bleeding as that will spread the mites and require the bird to be isolated.

Once the damaged skin is removed, cover the legs with petroleum jelly which will suffocate the mites.

This mite is most common in damp conditions, so ensure the run is not too damp and that there is suitable ventilation in the coop.

Lice
In chickens, lice eat skin debris and dead feathers. However, as they live at the base of the feather, they can cause severe irritation.

Lice are one of the easiest pests to treat, and you should check the chicken regularly for them. Hold your chicken by the legs and check the vent, where the eggs and droppings come from. This is one of the most common sites to find lice.

In a healthy chicken, the skin will look clear, and the feathers will be dry around the vent. With a lice infestation, the skin can redden, and you can see white crystals, like sugar, around the feathers.

Treat the chicken with lice powder if you find any signs of this pest. Another treatment is to rub in cold ash from a wood fire, which does a similar job.

Coccidiosis
This is rarely a problem for the backyard farmer, being more common in intensive chicken farming. However, if you or a neighbor have bought ex-battery hens, then they could carry Coccidia and introduce it to your flock.

Good hygiene prevents most cases of this, and its symptoms include blood in the droppings, extreme weight loss, lethargy, sudden death and reduced laying capacity.

Cockerel Behaviour
Most backyard farmers are not going to own a cockerel because they are not productive birds, though they are good for eating and producing fertile eggs. By nature, cockerels are very protective of the female birds and will defend them against predators and humans alike!

If you have multiple cockerels and they do not feel there are enough females in the flock, then they will fight. Some breeds are extremely violent, and the males will fight to the death.

Unless you are an experienced chicken owner if you want a male chicken you need to choose a more docile breed, rather than one of the more violent breeds. Though do remember that you may not be able to keep a cockerel due to your HOA regulations or other laws.

Like the hens, you should handle the cockerel to tame it and get it used to you. You'll be surprised how much damage a cockerel can do if it decides to attack you! Cockerels have spurs on the inside of the legs (they are less pronounced on female birds) which they use when fighting. Removing these is permitted in a few countries, but in most places, it is banned as it is unethical and mutilation. However, you can trim the spurs using claw clippers or file them down. This will help to prevent damage to your hens.

Crop Impaction and Stasis

These are both fatal illnesses if they are not treated, so you should keep an eye out for it. Crop stasis is also known as sour crop because the chicken's breath smells of sour milk.

Quite often you see the crop is full and not realize that the chicken has stopped eating. You can see the crop continue to grow as the chicken eats, but the food doesn't get through to the stomach.

In most cases, crop stasis isn't the primary illness and is a secondary condition caused by something else. Crop impaction results in the crop becoming red, hot and heavy, eventually making it hard for your chicken to stand upright. Watch your chickens eat and make sure they are really eating, not just pecking at the ground.

For these conditions the only treatment is surgery. If you catch crop stasis early enough, then it may be treatable without surgery before the impaction builds up.

Egg Peritonitis

This is most common in ex-battery hens and, unfortunately, does not bode well for the chicken.

As battery chickens are kept in harsh, crowded conditions where lighting is manipulated to maximize egg production the chicken has no time to recover from egg laying or to replace the calcium lost from their bones in the process.

Don't let this put you off rescuing ex-battery hens, however; they can have this disease when you get them but are showing no signs symptoms. Any chicken that has low calcium levels can suffer from this problem, which is why giving your chickens oyster shell grit is so important.

In this condition, the egg breaks inside the chicken, which then gives dangerous bacteria a perfect growing environment. If you catch this early, then the chicken can be treated with surgery or medical intervention. If in any doubts, take your chicken for a check-up as it is best caught while still treatable.

Scald
When a chicken has scald, the skin on its legs and particularly the feet go a bright red color, like it is scalded from being dipped in boiling water. This is a very painful for the chicken.

This condition is down to problems in the coop. This is caused by bedding that is too dirty with chicken poop. The high level of ammonia burns through the skin on the chicken's legs.

This is easily avoided by regularly cleaning the chicken coop and ensuring there is sufficient ventilation to remove the ammonia vapors.

Prolapse
Birds do not have a separate birth canal and anus, unlike mammals. The eggs and droppings come from different tubes on the inside and are delivered through the same vent, though obviously not together.

To prevent eggs being contaminated with droppings, there is a clever system on the inside where the tube that the egg comes out of (the oviduct) inverts until the egg is out.

In most birds, this works perfectly well, but when chickens are low in calcium or other vitamins, have a muscle weakness or has not had enough time to recover between eggs, the tube can come out, which is called a prolapse. This is easy to spot as you will see something red/pink at the vent.

If this gets worse or gets infected by droppings, then there is a good chance the chicken will not survive.

When your chickens are laying, keep an eye on them and take them to the vet if you see a prolapse. Although they can correct themselves, it is better to get treatment rather than leave it. The vet can also give your chicken a vitamin injection to help strengthen her. Immediately provide your chickens with a vitamin supplement in their water to ensure the entire flock has enough vitamins.

Marek's Disease

This unpleasant disease is most common in chickens under five months old. It is caused by an alpha-herpes virus which infects the immune system. This will cause nervous problems and even tumors.

Commercial chickens are vaccinated against this disease, and you can get your flock vaccinated too. However, the vaccine is only effective when the chickens are very young. You cannot vaccinate adult birds. Sebrights and Silkies are very prone to Marek's disease.

Indications of this disease include paralysis of either one wing or one leg, noticeable by a dropped wing. The head can be tilted, and the droppings can change. Marek's is spread through feather dust.

This disease, unfortunately, cannot be cured and the only solution is to euthanize and destroy the bird. If you are buying birds from a breeder, check to see if they have been vaccinated and ask for them to be done if they are not. The first injection is given at one day old, with a second at two weeks old. If you are breeding chickens, then be prepared, getting the vaccination ordered before the hatching date.

Mycoplasma Gallisepticum

This s a very infectious disease that causes breathing difficulties, which then takes weeks of antibiotics to clear up. It is, though, very hard to eradicate completely and can reappear again in the future.

Chickens who have this disease can have a clear discharge from their eyes and nose and may be sneezing. The eyes will often puff up, and the head can look swollen because the sinuses film up. Chickens can be lethargic, lose weight and even appear depressed. They can be dehydrated, which results in the comb falling over to one side.

Often these signs are ignored and if the disease progresses too far the chicken will not respond to treatment and will need to be euthanized. Caught early, it is relatively easy to treat with a course of antibiotics, though ensure the chickens finish the course of antibiotics otherwise the disease will quickly return.

Osteomyelitis

This is an inflammation of the bone, often found in the keel. Check this area for scabs or a wound, which indicates this problem. Your chicken may also develop a fever and lose weight. Regularly checking your chickens will help you catch this early on, which makes it much easier to treat.

Aspergillosis
This is a fungal disease caused by the commonly found fungus Aspergillus fumigatus. As this fungus is almost everywhere, most birds are exposed to it. In particular, it grows on moldy grains and seeds.

This means that it is very important you keep food clean and fresh as well as clearing up any scratch that your chickens do not eat. Good run hygiene will keep the risk of this disease to a minimum.

It is rare for this to be a primary infection, more commonly being a secondary infection in a bird already weakened by another illness. Therefore, if you have ill chickens, you should keep a close eye on them for signs of this disease.

As the fungus will grow in their air sacs, they can lose their voice or their tone of voice can change. In more advanced infections, the tail can bob up and down as they walk. Lethargy can be another sign, as can a slowness to come for food. They are also going to struggle for breath.

This is serious in adult birds and fatal in chicks. Therefore you should seek treatment at the very first signs.

You can prevent this disease by keeping the coop clean and making sure there is plenty if ventilation. Keep the run clean and make sure whatever material you use in it is regularly changed and does not get moldy.

In rare cases, this can infect people, particularly the young and elderly or people who are being treated with strong drugs, such as cancer therapies. It is best they avoid going near your chickens or the run if this disease is present.

Salmonellosis
Salmonella is a bacterium found in the intestines of virtually all birds, though is much more prevalent in chickens than other bird species. In normal cases, this will not affect your chicken. However, if you eat undercooked meat or eggs from these birds there is a high risk if you or your family contracting it.

Practicing good hand hygiene after handling the chickens, their eggs or bedding will help prevent the spread of this bacterium.

It's best to avoid feeding the chickens kitchen scraps, which was their traditional fayre. In a lot of countries, this is illegal to prevent the spread of disease.

If you have any concerns about the level of salmonella in your chickens, then you can send a fecal sample away to a laboratory for testing. If your chickens have salmonellosis, then they can be treated with antibiotics.

Chicken Health Checks

It is very easy to spot a sick chicken; they stop eating, are lethargic and will often be hunched up looking very sorry for themselves. Worryingly, by the time it gets to this stage the illness will have progressed quite far. Like many prey animals, chickens hide their illnesses well because weakness invites attack both from other chickens and predators.

Regularly spending time with your chickens helps you to understand their healthy personalities so that you can spot the slight variations in behavior that indicate illness. Quick action can often save a chicken's life.

Daily Checks

When you let your chickens out of their coop in the morning, just watch them to see if they behave normally. They should run out for their breakfast in their usual order, though don't interfere in this and let them eat as you could disturb their social hierarchy.

Picture © Matt Seppings

As they feed, just make sure they are not limping, there are no injuries or any trailing wings. If any chickens are bleeding, then isolate the bird as chickens are attracted to blood and will attack any bird that is bleeding.

Any bird that isn't eating or drinking properly should be examined more closely to see what is wrong and taken to the vet if necessary. You will get used to what is normal for each chicken after a week or two, and you will learn to quickly spot any problems.

Other things to look out for include:

- Glossy and unbroken feather, unless the chicken is molting
- Purple combs – indications of circulatory or heart problems
- Pale combs – natural in young or molting birds but an indication of problems in adult birds. The comb should be firm and red
- Watery eyes, difficulty breathing, snuffling, all of which indicate respiratory problems.
- A bulging crop – indicates a blockage – it should be empty before the chicken eats
- Droppings should be firm in consistency, dark brown and white. About one in ten will naturally be sloppy and foamy.

You can think this is a lot to do, but all of these observations are done while you are doing your daily jobs in the run and coop. Initially, this may take a little bit of an effort, but once you are used to doing it, you will be surprised how little effort these observations take.

Is Your Hen Sick or Just Broody?
You pop your head into the coop and see a hen sat in the nest box. It looks a bit dazed and has no interested in coming out for food. She isn't very alert; her comb is pale, and she has puffed up all her feathers.

So, is she broody or should you be straight on to the vet to get the bird checked out?

Picture © Pava

A broody hen will usually growl or squeal when approached and her breast is going to feel hot. She may even pull out some of her feathers. If you pick her up, she should be too preoccupied to peck, though some will. Remove her from the nest box and put her on the ground, at which point she should race over to the food, act normally and then return to her nest box. That's a broody hen.

If when you put the hen on the floor, she doesn't move or shuffles away to hide, isolating herself from the other hens, then you have a sick hen.

In-Depth Inspections
About once a week it is worth doing a more detailed inspection of your chickens. This will not take a long time to do but could make the difference between losing a hen and saving her.

Firstly, you need to catch each of your chickens. This is harder to do than you may think if you haven't tamed your birds. Whatever you do, do not chase them as that will stress them. Either use treats to call them over to you, wait until they are roosting or use a large landing net, as used in fishing.

With each hen, you need to check the following:

- Weight – regular handling of your chickens will help you to know whether they are gaining or losing weight. Feel for the breastbone, which will be sharp if the chicken is underweight. If the breastbone is under a layer of fat, then you need to put your chicken on a diet as fat hens don't lay eggs!
- Injuries / Swellings – Check over the chicken for any lumps or cuts
- Ears / Nose / Eyes / Vent – these should be clean and with no discharge. Check there is no rattling noise when the hen breathes
- Crop – should be empty if the hen hasn't eaten. If it feels firm and the hen hasn't eaten recently then it is impacted, and you need to take her to a vet. If her breath smells of sour milk, then she has a fungal infection and needs treatment
- Skin – use your fingers to gently part the feathers to check for lice and mites, particularly around the vent, on top of the head and under the wings. If you see white clumps stuck on the shaft of feathers that is usually lice eggs. If you see a greasy black mass, then that is the eggs of the Northern fowl mite. You will need to treat the whole flock and the coop to get rid of the mites

- Legs/Feet – scaly leg mite is indicated by raised scales. Swelling on the bottom of the foot indicated an infected injury, known as bumblefoot. Both require immediate treatment. When mud is stuck to their feet, soak it off rather than pull it off as you risk harming the chicken
- Trimming Nails – carefully give overgrown nails a trim or file as required, though usually nails will naturally be kept trimmed through activity

Once you've given the hen her health check, you need to give her a treat as this will train her to enjoy the checks, and over time it will get easier for you!

Potential Predators

We have talked about protecting your chickens from predators, which is vital. Depending on where you live you may be subject to one or more of these predators. Of course, this isn't a definitive list as there are plenty more specific to certain areas and countries, e.g. mountain lions, mink, cougar and more.

- Foxes – the main predator in the UK and Europe. A very opportunistic predator that will take advantage of any weakness in your security, hence why your run and coop needs to be well built. Common in urban areas as well as rural locations
- Badgers – very strong animals that will tear weakened panels off a chicken coop. Usually, they will only kill a single bird at a time whereas a fox will kill multiple birds, often the whole flock.
- Mink / Stoats / Weasels – all three of these will kill chickens, but they are rarer, usually seen in more rural areas.
- Domestic Cats – these, surprisingly, will rarely bother adult hens but should not be trusted around chicks and be watched carefully around small bantam chickens. Your cat may be fine around your chickens, but other cats, particularly feral ones, that live in the area may not be so chicken friendly.
- Rats/Mice – more of a problem because they spread disease, but rats will steal eggs, gnaw at woodwork and even at the chicken's feet. Rats are very discreet and harder to spot than mice. You will only ever have a problem with one or the other, not both at the same time as they do not coexist.

Vaccinating Chickens

This is a good way for you to prevent your flock contracting specific diseases, though it shouldn't be a substitute for good care and cleanliness.

A lot of backyard farmers will not vaccinate their flock because of the cost and availability of vaccines. Often, they need to be bought in large quantities because they are aimed at commercial farmers, or involve additional vet fees which make them uneconomical. Although it may be unnecessary to vaccinate your hens and they will be healthy throughout their life. However, do you want to run the risk of your flock being wiped out because you didn't spend the money protecting them from an easily preventable disease?

If you buy your flock from one breeder all at once, then your chickens are unlikely to be carriers of disease. If you buy from poultry shows, you have no idea what your chickens have been exposed to. Birds, like humans, can carry diseases without showing any symptoms. Likewise, if you buy your flock from a variety of sources, you have no idea whether any birds carry a disease.

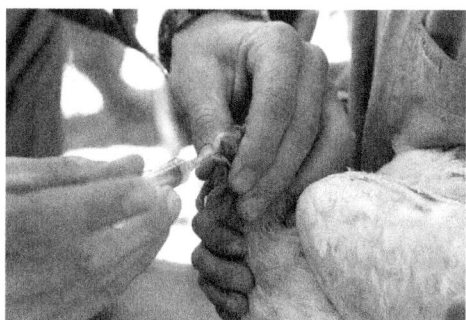

It is not recommended that you mix birds that have been vaccinated with those that have not, particularly when young. As most hybrid chickens are vaccinated, people often mix vaccinated hybrids with non-vaccinated pure bred birds. Be aware that vaccinated birds can still carry a disease like Marek's, without displaying any symptoms, that they can then pass to unvaccinated birds.

Vaccination must be carrtied out when the chickens are young, with most being done in the first few weeks. If you are hatching your birds, then obviously you can vaccinate them yourself. If you are buying birds from a breeder, you can sometimes request they are vaccinated for you. Not all breeders will be prepared to do this, and most will expect you to pay for the vaccination.

It is a tough choice as to whether to vaccinate your chickens because you have to consider the cost, not only of the vaccine but of the equipment to deliver it and possible vet fees if you cannot do it yourself.

If you have a static flock and are not introducing new birds, then the risk of infection is minimal, particularly if all your birds come from one, reputable breeder. Remember that some breeds, such as Silkie and Sebright are high risk to Marek's disease and therefore vaccination is strongly recommended.

When it comes to expanding your flock, it is probably best they go in a new area, so you do not mix birds and risk the spread of disease. Good husbandry and good cleanliness will keep the risk of disease down to a minimum. So long as you do not introduce new birds to the flock and wild birds have no access to the run or coop your risk of disease will be minimal. If you regularly visit other chickens, then you do run the risk of bringing disease into the flock yourself on your shoes or clothing. This includes children's petting farms, zoos and other locations that have chickens.

Vaccination is a personal decision, and you need to decide the risk to your flock, whether you will expand your flock and whether you can afford the expense.

How Stress Affects Chickens

Chickens are quite delicate animals and respond very negatively to any stress in their lives. Not only does it significantly contribute to illness, but they also become more prone to anti-social behavior such as egg eating, comb/vent pecking, and feather pecking.

Intensive farming is very stressful for chickens, hence the high mortality rate. However, chickens living in someone's back yard have a luxury life. They are subjected to a lot less stress and are much healthier, providing they have a big enough run and are let out of their coop daily. Saying that home kept chickens can still be subjected to stress.

It is best to avoid causing any stress for your chickens, and with a bit of thought, this isn't too difficult to do. Some years ago, scientists researched the causes of stress in chickens by measuring the levels of the corticosterone hormone (associated with stress) in a variety of different situations.

Chickens suffer from stress in many different situations, some which you may be quite surprised about, but others make sense. The most common stress inducing situations include:

- Handling – being a prey animal, being picked up does not have a good association with them as it usually involves being eaten! Chasing your birds around to catch them and handle them is going to cause lots of stress. You need to regularly handle your chickens (while doing health checks) and get them used to being picked up. It's best to catch your bird in their hen house quickly. It's important you hold the wings firmly and then you hold the chicken with your right hand under her, holding the breast in the palm. The fingers of your right hand hold the tops of her legs, with her head underneath your arm, so she looks behind you. You then have your left hand free to examine her, clip nails or whatever else you need to do.
- Introducing New Chickens – there is a social hierarchy within your chicken flock, and every bird has its place. When you introduce new chickens, suddenly this pecking order is upset, and the birds are now no longer sure where they stand, who should eat first, who should roost highest and so on. Naturally, this is going to cause fights as the chicken's jockey for position once more. Introducing a new chicken will cause these fights as it finds its position within the flock's social hierarchy, but it should settle down in a few days. You must keep a close eye on your chickens during this time and remove any injured birds. You should also never introduce a more timid breed into a flock of more aggressive birds as it will get bullied.
- Lack of Resources – this is an easy stress to avoid, but chickens will get stressed if there is a lack of resource, whether this is food, water, dust bath, perching space, nesting space, grit or something else. Ensure water containers do not leak and do not freeze, ensure plenty of food for your birds and plenty of space. If they have all the resources they need, then they are not going to have anything to stress about. Lack of resources will cause fighting amongst your flock and can end up with injured birds.
- Extreme Heat – as chickens don't sweat, heat is difficult for them to handle. They either drink more water (needs to be cool so remember to put their water container in the shade) or they pant, losing heat as they exhale. Chickens find it easier to handle cold as they can fluff up their feathers, huddle together and eat more. Make sure their coop has plenty of ventilation and does not get too

hot, as excessive heat is going to cause stress. There must be a shaded space in their run where they can get out of the sun.
- Moving/New Environments – chickens do not like change. Moving them to a new run or new coop can cause stress. Taking the bird to a vet or a show will also cause stress. Bizarrely enough, a snow fall will cause stress because the chickens think everything has changed because it is white and doesn't look like normal.
- Egg Laying – a natural process for chickens but it does cause them stress. A nice quiet, dark nest box will help to reduce their stress. Leave the birds alone while they are nesting as disturbing them causes stress too.
- Predators – not overly surprising that this causes stress, but having predators around the coop or run is going to cause stress. They will also be stressed if they know a predator is lurking in the bushes near their run, so keep the area around their coop clear so predators must show themselves or keep their distance.

Stress is that serious in chickens that it can be fatal, so do your best to minimize it wherever possible, and you will have a happy, healthy flock that are good layers. As you can see, almost anything causes a chicken stress, so you need to look after them!

Red Mites

Red mites are one of the most common problems chicken owners face. They live in cracks in the coop, particularly at the ends of perches (why is why you need them to be removable). At night, they come out of hiding and feed on your birds.

These are quite hard to spot unless you know what you are looking for. They start life as small, white/gray mites that eventually, after feeding, turn red and grow to a tiny 0.7mm in size!

One of the indicators that red mites are present is a gray dust, often around the perch ends, that looks like cigarette ash. This pest is most active at warmer times of the year, typically from May to October and go dormant during winter. Their life cycle is just a week, so they multiply and can become a major problem very quickly.

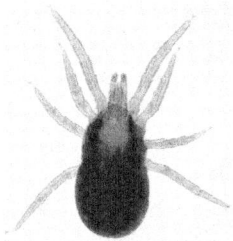

Picture © Daktaridudu

It is best that you are proactive regarding red mites as prevention is far better than a cure. Every week, when you are cleaning the hen house, check for red mite. Once a month in the summer, do a deep clean of the hen house, using a preventative treatment to stop them from getting a foothold in your coop.

Most chicken owners will not notice red mite until it is a major infestation and the hens refuse to go into their coop at night. They can end up becoming anemic, the combs turn a paler red, they stop laying, and you can even find red blood stains on eggs which are squashed mites. In the worst cases, your chickens will start to die.

Red mites hide in the day but can be seen when you lift away the perches. An easy way to check is to visit the coop at night when the chickens are roosting and run a tissue along the bottom of a perch birds are roosting on. If it ends up with blood on it, then that's red mites you've just squashed.

Getting rid of them completely is very difficult because they are very good at hiding. It is far better to try to prevent them than trying to keep on top of them after you have an infestation.

If you do find red mites, then you need to do a major clean of the coop. It is going to take you a couple of hours, plus you need to do a further clean which takes around an hour every five days for a couple of weeks.

Two of the best treatments for red mite are Poultry Shield and Diatom. You can also buy a red mite powder to treat the birds before their roost to help protect them from the mites.

Remove all the birds from the hen house and strip out as much as you can. The bedding needs to go straight into garbage bags, sealed and disposed of as red mites can live around six months without feeding.

Mix up enough Poultry Shield to treat the entire coop and water everything with it. You need to get this mixture into all the cracks, down the edges of the coop and particularly at the perch ends.

Leave this to soak for a quarter of an hour and then return to the coop. The red mites should be fleeing their hidey holes now so pour more poultry shield on them and their hiding places.

Leave it again for the same length of time and then hose out the coop, paying particular attention to the nooks and crannies. A pressure washer is best for this because it will blast any hiding mites out of the house.

Leave the house to dry before reassembling it, and then use some Diatom which will help kill off any mites that somehow survived the cleaning.

Once the house has dried and aired, you can put new bedding in and allow the chickens to return.

Dust the ends of the perches/nest boxes and where ever else you found concentrations of red mites when cleaning. Repeat this within a week so that the mites don't have a chance to lay more eggs and spread again. You can find red mites setting up home under the felt roof of the coop, which makes it very hard to get rid of them without removing the felt.

Dust the chickens down with Red Mite Powder, which will help give them some respite at night from the mites when they are most active.

Bird Flu and Chickens

Avian Influenza or Bird Flu is something that can potentially infect your birds so you should be aware of it. Firstly, at the time of writing the disease does not transfer between humans and chickens and vice versa.

There are two types of bird flu; low pathogenicity (LPAI) or high pathogenicity (HPAI). Which one it is depends on how severe the illness is that it causes. HPAI is very contagious and will quickly pass from bird to bird. It is typically deadly, and the presence of it in your flock will require the entire flock to be culled and destroyed.

Chickens get infected by direct contact with infected chickens, infected waterfowl or feed/water that has been infected. If your birds are kept in a run and coop where they cannot get near other birds, then they should be pretty safe, but in an outbreak, you are advised to keep your birds inside.

You need to be aware of bird flu because if there is an outbreak, you should avoid going to places where there are other birds, such as zoos, other chicken keepers, parks or lakes where there are wild birds and so on. In fact, you should probably even stop feeding birds in your garden, if that is something you do, to keep your flock safe.

Avian influenza is quite rare, but there are occasional outbreaks which governments work very hard to contain as it could decimate the poultry industry if it were not swiftly dealt with.

If bird flu is a risk, then bring in your free range chickens, keep your chickens in their coop or small run (ideally cover it so droppings from other birds cannot get into the run) and avoid visiting any locations where there are other chickens or waterfowl. With reasonable bio-security and a bit of common sense, you will avoid a bird flu infection. More than anything else, bird flu is an inconvenience as you must keep your birds inside and avoid even visiting places where there is a risk of infection. You can bring the disease into your chickens on your shoes or clothing.

You are unlikely to see this disease in your chickens, but if there is an outbreak, you must take precautions. You will typically find out about outbreaks on the news, from other chicken owners or on poultry forums or groups online.

WHY CHICKENS STOP LAYING

Eggs are the main reason most people keep chickens and are a very healthy food, in moderation. However, there is going to come a time when you wonder why egg production has decreased or even stopped.

There are several factors at play in egg laying and a change in any one could well cause your hens to slow down or stop laying. Practicing good management techniques will minimize these problems, but your chickens naturally lay fewer eggs at certain times of the year. This is good for them because it gives their bodies a rest so they can build up protein and vitamin levels so they can lay well the following year.

Here are some of the main factors that influence egg laying.

Decreasing Day Length
Egg laying is tied to the length of the day, and as the amount of daylight reduces, so egg laying reduces and stops. The reason is quite simple, less daylight means winter, which means fewer resources and the chickens need to conserve their energy to survive the cold winter to lay eggs next year. Let's not forget that in the wild, chicks would have less of a chance of

survival in winter.

Hens usually need fourteen hours of daylight to produce eggs, but as daylight goes below twelve hours a day so egg production drops noticeably and may even stop if the days get too short, as they would in a more northern latitude.

In a commercial environment, this is overcome by using artificial lighting to trick the hens into maximum production. However, this is not good for the hen's long term and is not recommended for the home farmer. Let your hens work in their natural cycle, and they will have longer, more productive lives as healthy hens.

Breed
You should also consider the breed that you have chosen to keep too. Some breeds are prolific layers and will lay close to an egg a day whereas others may only lay one every few days. It is quite easy to forget this and to wonder why you aren't getting as many eggs as your friend across the stress only to realize later it's purely down to the difference in the breed.

Poor Nutrition
Feed your hens too much, and they will get fat, which makes them stop producing eggs. Feed your chickens too little, and they will get too thin and not produce any eggs. See the problem here?

Producing an egg takes a lot of energy from the chicken and depletes them of vital vitamins and minerals. They should have a good quality feed which is well-balanced and high in protein. Typically, you will use a layer's pelleted food.

As well as this, give your chickens oyster shell grit to help them absorb more calcium and a multivitamin in their water to replenish other vital minerals. Poor nutrition leads to a wide variety of other problems too such as a prolapse.

As well as a good quality food, your chickens will need fresh, clean water, something that is often overlooked. This water must be clear of chicken poop and should be kept in a shady place in summer with the ice broken in the winter. A bird can drink more than you think in a day, and without water, they will quickly stop laying and even die.

Molting
Chickens generally molt towards the end of summer, when the days start to

get shorter. For a hen, molting is a very stressful time that requires a lot of energy and specifically protein. A chicken cannot molt and keep laying so when this process starts, egg production stops. It is not uncommon to see an entire flock stop laying and molt all at once.

Molting can take anywhere from six to twelve weeks, and once it is completed, the hen will start to lay again. During this time the hen's body has a rest from laying so her reproductive tract can recover and her feathers are restored.

Broodiness
This problem affects some breeds more than others, though many of the hybrid strains have had this tendency bred out of them. Broodiness is associated with hatching eggs, so you will see your chicken stop laying and start to gather together eggs and sit on them. She will sit in the nest box most of the time, which will stop the other hens from laying their eggs. If you collect eggs every day, then this will help to prevent broodiness as the hen cannot build a clutch. Breed selection is also an influence in how likely this is.

You can leave your chicken to raise her eggs naturally, but be aware that she will not be laying eggs for a few months. Commercially, this is an economic nightmare, but for the home breeder, this is by far the best way to raise chickens as your hens are much more knowledgeable about it than you.

Age
This is a significant contributing factor in egg production. In her first year, chickens will lay more eggs than in later years. She will continue to lay well in her second year, but you will notice fewer eggs than in the first. After this, you will see a decrease in the quantity of eggs laid each year. You will also see a decline in quality of both the albumen and the egg shell.

Commercially, chickens are kept for no more than two growing seasons. Home farmers tend to keep them longer, often to the end of their life. Some people will keep the chicken while it is a productive layer and then when egg production drops too far serve it for dinner.

Disease
An infected hen will often see a decrease in egg production. Diseases such as infectious bronchitis, fatty liver syndrome and insecticide poisoning will all cause egg production to drop. If other signs point to disease being the cause, then your hen needs to go to the vet.

Stress

Chickens love their routine and get very stressed if their habits are changed. Any number of factors can stress a chicken, which will cause them to either stop laying completely or lay fewer eggs. Some of the more common stress factors include:

- Temperature Too High or Low – chickens cannot sweat and so get stressed if they cannot cool down and stop laying. They need cool water and shade to manage their temperature as well as the ability to get inside out of the cold.
- Damp / Draughts – something else chickens find stressful. Make sure your hen house is draught free and does not get damp to keep them happy. Also, make sure the run isn't too damp or muddy as they will not like that.
- Sufficient Food – chickens need enough food to lay eggs and to keep themselves warm in winter. Expect their food intake to increase in winter and decrease in summer.
- Moving / Handling – both of these will cause stress, but if you train your birds to be handled, then they will learn to find it less stressful. Moving them to a new location or taking them to shows will also cause stress.
- Introducing New Birds – this upsets the social hierarchy of the flock which will also cause stress. Expect egg production to decrease if you introduce new birds into the flock.
- Ticks / Lice / Mites – all of these cause stress, and so you should keep ticks to a minimum through good management practices.
- Fear – as prey birds, chickens are scared of almost everything. Things like dogs barking, larger animals, predators, young children, birds of prey, loud noises and vehicles can scare the hens and decrease their egg production.
- Egg Eating – chickens will eat broken eggs and once they have acquired a taste for eggs, will try to break eggs themselves to get at the inside. This is a difficult one to solve, but if you collect eggs daily, remove any broken eggs immediately and take what preventative measures you can, then you should hopefully avoid this problem.
- Lack of Nest Boxes – you need, roughly, one nest box for every three hens because if there isn't enough nesting space, then your hens will either lay eggs on the floor or not lay at all.
- Lost Eggs – some chickens prefer not to use a nest box to lay in, instead choosing to lay anywhere they want. Keep an eye on the

run, particularly under bushes to find any eggs not laid inside the nesting boxes.

Looking after your hens is very important for regular, healthy egg production. Taking good care of them will ensure the egg production season is extended and you may even get some fresh eggs during winter.

Safely Storing & Using Fresh Eggs

Fresh eggs are absolutely wonderful, and once you have had some, you will find it very difficult to go back to the bland supermarket eggs. Let's take a moment to talk about how to collect and store your precious eggs.

Collecting Eggs
Eggs are usually collected in the morning. When you let your hens out of their coop, check the nest boxes for fresh eggs and remove any that you find. When you put your chickens to bed at night, check again for eggs and remove any you find. Leaving them in place encourages broodiness and egg eating.

Cleaning and Storing
Eggs have a natural coating, known as a 'bloom,' which protects the egg from bacteria. When you collect eggs, do not wash them as this removes the protective bloom. Wipe it with a rough, dry cloth.

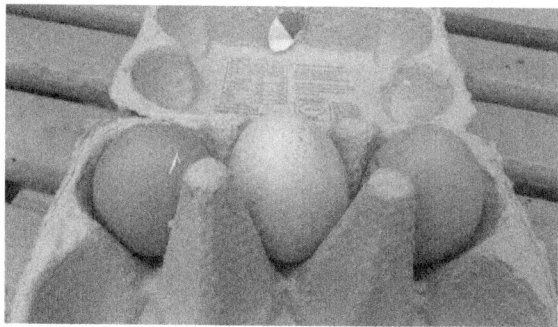

However, a really dirty egg will need scrubbing with a vegetable brush in warm water. Avoid using cold water as this encourage bacteria and makes the egg shrink in its shell. If there is a little manure on the egg, then you can carefully wipe the spot with a dry cloth. Use eggs that have been cleaned first as they are not protected by the 'bloom' and will not be good to eat for as long.

Allow the eggs to dry thoroughly in the air before you put them away. Store them in dated cartons, so you know which ones to use first. They are better on a shelf in your refrigerator rather than on the door as they do not get jostled around as much. If you do not have enough eggs to fill a carton then you can write, carefully, the date it was collected on each egg.

As a tip, when cooking fresh eggs, their shells stick to the inside rather than cleanly peel, whereas eggs that are a week or so old peel much easier.

Storing Eggs for Incubation

Hens will collect eggs together for a few days, then sit on them and become broody. If you want to incubate and hatch eggs, then you should mimic this natural behavior. Collect eggs for a few days and store them somewhere out of direct sunlight at a temperature of 55F. Humidity needs to be low enough that the eggs do not get moldy but high enough that the eggs don't dry out, so don't refrigerate them!

After six days, no matter how well you store them, their hatchability will decrease. Some people reckon that wrapping the eggs in plastic wrap will keep the eggs viable for an additional three weeks. Store your eggs in used, clean egg cartons with the small end down, which keeps the yolk centered. Storing them the other way up means the yolk sinks down into the fat end of the egg which can cause problems with hatching.

When storing eggs for longer than six days, you should turn the eggs every couple of days to stop the yolk sticking to the shell.

Anyone planning to hatch eggs should save the best quality eggs for hatching and store them very carefully. Hatching your own eggs is a great way for you to expand your flock, and we'll discuss this later.

Raising Chickens for Meat

Raising birds for the table, or broilers as they are known, is less common amongst home farmers, but a lot of people still do this. What puts people off is killing and butchering the bird, but you can take them to a butcher who will slaughter them for you.

The process for raising chickens solely for meat is different when compared to raising laying birds. Firstly, you are likely to choose different breeds, and secondly, you feed them a different food, and you do not keep them as long. Make sure though that you do not get attached to the chickens and avoid letting children get involved with meat birds as they will get upset when they are served up for dinner.

Meat birds need the same level of care as laying birds, which you are well aware of. For male meat birds, you can keep them in the pen with female birds until about 14 to 16 weeks of age, feeding them the same diet. At around 16 weeks, layers are moved on to a special food.

At this point, you can leave the cockerels in with the female birds to eat the same diet. This is easier for a lot of people as you only need the one coop and run, but it does mean the birds grow slower. However, they do have a good flavor, but you can improve that flavor by isolating the bird three days before slaughter and feeding it corn. Birds kept like this will be ready at 22 weeks of age, though you can leave them longer to put on more weight. Just watch the birds closely as sometimes the cocks will fight amongst each other or bully other hens.

The other method is to move the meat birds out of the run, away from the egg layers. These birds are then fed finisher pellets which are designed to help them put on weight quicker so you can butcher it. As before, you should move them to a corn based diet three or so days before killing them. You will need a second run and coop to keep your meat birds in, which not everyone has the space or money to do.

If you can separate your meat birds, then there is a feeding regime you can use which will help them to grow to broiler size quicker. At 3 to 4 weeks old the meat bird is moved from starter or chick crumb to grower's pellets. At seven weeks, old the bird can be moved to finisher pellets. In about 8 to 10 weeks, the bird will be of a size suitable for eating. When feeding your birds this diet they should have plenty of food and water available, so top it up a couple of times a day!

Meat birds, like the egg layers, need to be kept under heat for a minimum of four weeks after hatching, though a broody hen will do this job for you. Particularly with broilers, you need to ensure they are warm and clean so you don't have any losses in the first few weeks.

Choosing A Meat Chicken

Some people will get a specific breed as a meat bird while other people will just use any male birds that hatch as broilers. Male birds are very hard to rehome, so they are either culled when young or raised for the table. There are several breeds which are well suited as meat birds, which we discussed earlier on in the book, here. Be aware that bantam birds, although they would taste okay, are a poor choice of meat bird due to their size.

Which bird you choose is entirely up to you. Whether you go for a hybrid, pure-bred or dual purpose bird is entirely up to you. Be aware that some of the modern meat breeds, particularly those that are used commercially, grow very quickly and can struggle to move around to enjoy the outside world. Meat bred birds are ready for the table very much quicker than dual-purpose birds. It is not uncommon for a meat breed to be ready for the table at a weight of close to 4lbs within eight weeks.

The free-range chickens you buy from the supermarket will be one of the rapidly growing meat specific breeds. They will appreciate the space and care of being raised at home, but you can comfort yourself with the knowledge that you have given the birds a good life before eating them.

Most people will tell you that home raised chickens taste far better than anything bought in the shop. Choose a breed that you like the sound of, have the space for and can get hold of easily enough. There are a wide variety of different breeds available, and it is, at the end of a day, a personal choice as to which you choose.

Killing the Chicken

This is probably the most unpleasant part of keeping a meat bird and the one thing that puts many people off raising chickens purely for meat. Sometimes a bird needs to be culled when it is ill, which a vet will do for you, but they can't provide any help with getting a bird ready for your table.

When a bird is suffering, and there is no chance of recovery, then euthanasia is the best option. It is much kinder to kill the chicken rather than allow it to continue to suffer. Experienced keepers can make this judgment themselves, but newer keepers will want a vet's opinion.

Killing a bird for meat or euthanasia follows the same process. If you are not happy doing this yourself, then take your birds to a butcher. It is better the job is done right than you cause the chicken pain and suffering by doing it wrong. One of the best places to learn this technique is to talk to another chicken owner and get them to show you how to do it when they kill one of their hens.

Neck Dislocation Method
You are aiming to minimize suffering to the bird so it becomes unconscious as quickly as possible. When this method is done correctly, as the neck is dislocated, the chicken immediately becomes unconscious.

This method is best learned from an experienced chicken keeper as it is quite hard to describe it clearly. Inexperienced keepers will find it easier to do on small birds, but it will be harder on larger birds such as heavier ducks.

Firstly, you need to catch the bird and keep it calm. This is best to do in the evening when it is calm and roosting. Take the bird away somewhere that it won't stress the other birds if it makes any noise. A shed or garage is ideal.

In your weakest hand, hold the chicken firmly by the legs. Ensure the legs are held together tightly and make sure that if it is a male bird, the spurs can't hurt you.

Rest the bird's chest on your thigh, which supports the weight. The bird will be upside down with its head being the lowest point and its feet being the highest.

Spread the first two fingers of your strongest hand. The back of the bird's head is held tightly between these fingers, with your thumb underneath the beak. Tilt the head back a little.

Using a firm action, you pull the neck sharply downwards, pulling the head backward as you do. Your knuckles are pressed into the vertebrae of the bird which bends the head back while stretching the neck.

There is going to be a lot of leg kicking and wing flapping, which is perfectly normal. It can happen a few seconds after you have dislocated its neck and is a reaction of the nervous system after the bird dies. Providing the neck is dislocated, the bird is in no pain.

When done correctly, you will feel the neck stretch and the head move down. Running your fingers down the vertebrae will enable you to feel the gap of the dislocation.

Before plucking and dressing the bird, you need to hang the bird upside down, by its feet, to allow the blood to run to the neck. You don't need to do this if you are euthanizing the bird, it only needs doing if you are eating it.

Picture © Superchilum

Put a container under the bird and then cut a slit in the gap between the vertebrae, severing the main artery and letting the bird bleed out. Putting a black garbage bag into the container can help reduce blood splatter and potential mess.

Plucking a Chicken

Once you have killed and bled your chicken, the next job is to pluck it, which is a skill you definitely need to learn. When done wrong it is an inefficient and unpleasant job. When you do it right though, it is a quick and easy job.

The traditional plucking method involves a technique now called scalding, which involves plunging the carcass into boiling water before plucking.

You need to boil a large pan of water. It needs to be big enough to take the whole chicken. Wearing thick rubber gloves for protection, hold the hen by its feet and immerse it in the boiling water. Use a wooden spoon to keep the bird immersed in water and let it soak; older chickens will need more time in the water than younger birds.

The chicken needs to stay in the water for anywhere from 5 to 30 seconds, depending on the age and size of the bird. You are not trying to cook the bird but just loosen the feathers, so they come out easier. The heat expands the skin at the base of the feathers so you can pull the feathers out easier without damaging the bird.

Depending on your skill and the size of the bird the plucking can take anything up to half an hour. Be more careful with young birds as their skin is more delicate and could tear.

Picture © Thamizhpparithi Maari

Wear gloves while plucking to stop your hands getting sore and if the feathers become difficult to remove, soak the chicken in boiling water again. The first few you pluck will be difficult, but your skills will improve. Don't worry because if you do a poor job of plucking the bird, you can still eat it.

Working at a fairly fast pace, you grab a handful of feathers, close to the skin and pull them out with a rapid action. Try to throw them directly into a garbage bag, so there is less tidying up to do afterward. Working at speed means it is easier to get the feathers off. If there are more pinfeathers,, then it is harder to pluck the bird.

You may notice some light, almost hairy feathers left after plucking. These are scalded off using a naked flame such as a barbecue lighter.

Once the bird is plucked, it is time to start the butchering process.

Scalding a Chicken

Scalding really does make the plucking process much easier, so we will discuss it in a bit more detail.

You are going to need a thermometer so you can measure the temperature of the boiling water so this technique is most effective.

Heat your pan of water to between 145 and 150F before dunking the bird into the water by its feet. It needs to be deep enough that the small feathers on the back of the legs are immersed in the water.

Hold the hen in the water for 3 seconds, then jiggle it around up and down to get the hot water to the base of the feathers. Pull the chicken out, plunge it back in again for 3 seconds, jiggle it some more then remove it again.

Now test the feathers to see how easy it is to pull them out. Try pulling out a tail feather or a long wing feather. If the feather comes out easily, you have scalded the chicken perfectly.

If not, then dunk the bird a couple more times until the feathers come out easily. There is no magic number here, how many times it needs to be done depends on the size, age, and even the breed of the bird.

If during the plucking it becomes hard to pull out feathers, scald the bird again until the feathers come out easily again.

Butchering A Chicken

Butchering a chicken is the hardest part for most people and the other thing that puts people off keeping chickens for meat. You can butcher the chicken yourself, as per this quick guide here or you can take it to a butcher who will do the job for you. Many people prefer the latter simply because they consider it cleaner and don't have to get their hands dirty. Slaughtering and butchering a chicken isn't a job everyone is up to so don't let that put you off keeping chickens for meat when you can hire a butcher.

Picture © Bhaskaranaidu

Initially, you will find this slow and tricky, but as you get used to it so you will master the technique and it will become quicker to do.

Start by preparing your butchering area. You need lots of space, and it needs to be clean. Sharp knives are also required, blunt knives make the job hard and messy. You will also need two bowls, one for edible organs and one for waste material. Get a large bucket or cooler full of very cold water, which is used to keep the meat cool after butchering as well as keeping flies off it. If you are only preparing a single chicken, then you don't need this, but when butchering more than one, it is very helpful.

Start by rinsing the plucked chicken well under running water before cutting off the feet at the joint of the yellow and pink skin. The tips of the wings also need cutting off.

Cut the skin around the neck and then strip the skin from the neck. The neck is cut off and either discarded or used to make stock. The crop is also carefully cut out and also discarded.

The tail oil gland, which is a deep yellow color, is carefully cut out next. Then pinch the skin above the vent and pull it upwards, which helps prevent you from cutting into internal organs and ruining your chicken. You next make a hole just above the vent which is used to remove all the innards.

When you make the incision into the body cavity, watch out for any liquid coming out. If it does, then you must discard the whole chicken as it was sick and is therefore inedible.

Enlarge the cut you made with your fingers, careful as some fecal matter can escape from the vent at this point. If so, wash the bird immediately, ensuring none of the fecal matter gets into the body cavity. Also, give the work surface a good wash if this happens.

You now reach into the chicken and pull out the organs, which may take two or three goes. Be very careful not to pierce or break the gall bladder or intestines. The latter will still be attached to the vent so needs cutting out carefully without damaging it!

Separate out the edible organs, such as the heart and liver, from the inedible ones. You also need to scrape the lungs out of the body cavity because they are very close to the ribs. A knife or other tool can help here as can rinsing out the cavity to get rid of the scrapings.

Once the lungs are removed, give the bird a thorough washing before either cutting it into pieces, such as the wings, breasts and so on or leaving it whole. You then refrigerate the bird if you are using it soon or seal it well and freeze it.

Keeping Chickens for Beginners

The unwanted parts of the chicken can be disposed of. Burn them, bury them or compost them in a secure bin with a lot of straw.

Now you have butchered a chicken and can enjoy the fruits of your labor!

BREEDING YOUR CHICKENS

Once you start keeping your chickens, particularly if you get expensive or unusual breeds, you may decide that you want to breed them. This allows you to expand your flock easily and with the more unusual birds, make some extra money.

When you hatch eggs, you will typically get slightly more males than females, so you need to be prepared to either dispatch the males, keep them or fatten them up for the table. You will need at least one male bird to fertilize the eggs in order to hatch them. More on hatching in the next chapter. People who keep meat birds will be interested in male chickens as they must continuously replace birds as they eat them. It is worth looking around locally and finding someone like this as they could be a good outlet for perhaps unwanted male chickens.

Before you breed your chickens, you need to know whether you can keep a cockerel. Although some HOA's and allotments will allow you to keep chickens, they object to male birds because of the noise that they make. Local residents are not impressed at a 4 am wakeup by an over enthusiastic cockerel! Without a cockerel, you cannot breed your chickens and will need to buy fertile eggs if you want to hatch hens.

You also need to consider whether you have enough space for the new additions to your flock. You need to keep the baby chicks separate as mixing them with adults risks bullying and them catching a disease.

When breeding an actual breed, you need to be selective of the hen(s) and cockerel that breeds. You want to use good quality, healthy chickens so that good genetic traits are passed down to the chicks. A poor choice of breeding partner can mean weak or diluted breeds. If the hen has any deficiencies such as bowed legs or bent tails, then don't breed them as there is a good chance these problems will be passed on. To selectively breed your chickens, you need to isolate the male and introduce females for breeding. This way you have complete control of which hens breed, and so have a good chance of getting good genetic material passed on.

Ensure all breeding chickens are healthy, with clear eyes, red combs and are bright and alert. The cockerel will be attentive to his hens, and if not then there is a chance he is not feeling well. Check for lice, mites, and other infestations before you start breeding. Females can benefit from poultry saddles to prevent damage to the feathers on their backs.

During the breeding process, you want to ensure that your chickens have a good diet, with plenty of their food pellets and lots of greens. If you find that a lot of chicks are dying in their shells, then switch to breeder's pellets which have a different mix of vitamins and minerals. You will need to keep your breeding hens away from the rest of your flock, so you don't accidentally collect fertile eggs.

Think about what traits you want to breed into your chickens. Do you have one that is a particularly good egg layer? If so, breed it, and there is a good chance its chicks will also be good egg layers. Do you have a particularly good looking chicken? Again, breed it and the chances are the traits will be passed on. Breeding is not only an opportunity to expand your flock but also a chance to pass on desirable traits to a new generation of chickens.

You do not want to breed chickens that are poor layers, produce poor quality eggs or have a bad temper as those traits will be passed on.

More experienced breeders can even create their own cross breeds, breeding together desirable traits from two different types of chicken.

Those who exhibit chickens will want to breed their show winning chickens for continued success and profit.

Breeding can be a fun process, but to control the breeding and ensure you get the desirable traits you want, you need to control the cockerel's access to the hens. Not everyone has the space to separate out breeding birds from egg layers, but if you do, this can be a great way to expand your flock and breed in desirable traits.

HOW TO HATCH EGGS

Hatching eggs can be a very enjoyable part of keeping chickens, and children, in particular, will find this fascinating. Incubators are now much more affordable, and so more and more backyard farmers have started hatching their own chicks. Affordable incubators can easily hold 20 to 25 eggs, which is more than enough for the home garden.

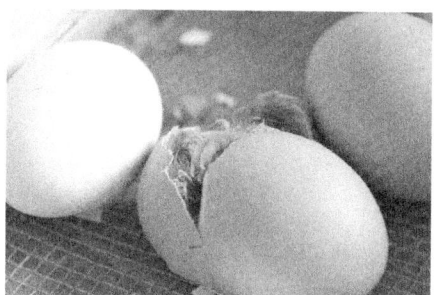

If you do not have a cockerel, then you can buy fertile eggs from breeders or online and use those in your incubator. Eggs from the supermarket and hens will not hatch; they must have been fertilized by a male bird to be hatched. Fresh eggs have more of a chance to hatch, so try to gather fertile eggs for a few days and then incubate them all together rather than incubating them a few at a time. This way they are more likely to all hatch at around the same time, which makes caring for the chickens easier.

Store fresh eggs in a cool, dark environment with the right level of humidity until you are ready to incubate them. You can use hens to sit on your eggs, though a lot of people prefer the control they have with an incubator and the fact it is more visible to their kids.

Remember though that you are going to end up with slightly more males than females so need to be prepared to deal with the male birds. Most birds cannot be sexed until they are eight weeks old, though there are auto-sexing breeds (chicks have different colors or markings depending on the sex). Even with these, you still must dispose of the new born chicks if you do not want to keep any male birds. Male birds are very hard to rehome unless you know someone who breeds chickens for meat, in which case they may take them. Most backyard keepers will not want male chickens.

When selecting eggs from your own flock to hatch, look for good quality eggs that have the classic egg shape to them.

Choosing an Incubator
Incubators come in two types, still and forced air. The main difference between the two is that the forced air incubator has a fan in it that circulates the air. This means that the temperature will be consistent everywhere inside the incubator. With a still air incubator, there is a temperature gradient, so it is often a few degrees warmer at the bottom than the top.

There are plenty of incubators available online. Click here to see those available in the USA and here for those in the UK.

For the beginner, the forced air incubator is the best choice as it is easier to work with and "get it right." If you buy one with an automatic humidity controller, which is a little bit more expensive, then you will find it much easier to hatch eggs successfully.

Incubating Eggs
Eggs are incubated for twenty one days at a temperature of 37.5C / 99.5F. As the eggs get closer to hatching, they will start to produce their own heat,

but the thermostat in the incubator will take this into account and keep the temperature consistent.

Picture © Santeri Viinamäki

Humidity levels need to be between 45% and 50%, and eggs need to be turned, by you, 180 degrees regularly. Some incubators have an automatic turning mechanism, but this feature adds to the cost of the incubator.

From a typical batch of eggs, you can expect anything from half to three-quarters of the eggs to hatch as not all eggs are fertile. It is very rare for an entire clutch to hatch.

Incubation Tips
Before using an incubator, it must be thoroughly sterilized with a disinfectant that is safe to use on an incubator. This kills off bacteria which will grow very quickly once you turn the heat on.

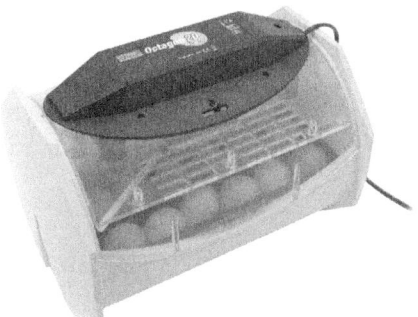

Available on Amazon

Before putting any eggs in, and after cleaning, plug your incubator in and leave it running for 24 hours to ensure it maintains a steady temperature of 37.5C. Water reservoirs need topping up daily to ensure that humidity levels are maintained.

It is always worth candling eggs after they have gone into an incubator. This will show you any damaged or cracked eggs that are not going to hatch.

Candling Eggs
This is done in the dark using a special light known as a candling torch. Once the eggs have been incubating for a week, you can use a candling torch which will show you blood vessels and the embryo inside the egg.

Unfortunately, you cannot determine whether an egg is fertile before you incubate it. Candling an egg before the end of the first week is not recommended as the additional heat can damage the embryo.

When candling you may see eggs that are cloudy, clear (infertile) or has blood rings, any of which means the egg will not hatch and should be discarded.

Tipping the egg very gently from side to side while candling will help you to see whether the egg is ok.

Most breeders will candle their eggs at seven days and again at 14 days to determine the viability of the eggs. Eggs that aren't viable need to be discarded.

At the broad end of the egg, you will see an empty area, known as the air sack. Between this and the part where the chick grows is a membrane. You will see this gradually increase in size throughout the incubation period until the chick breaks through into the air sack.

Hatching Chicks
A few hours after penetrating the air sack, the chick will "pip" the shell, but it can take anything up to another twelve hours or more for the chick to completely break out of the shell.

If the humidity level in the incubator is too high, then the chick can pip the shell and drown in fluids before their beak gets out of the shell.

If humidity is too low, the air sack is likely going to be too large with an undeveloped chick that can get stuck to shell or even be too weak to break out of it.

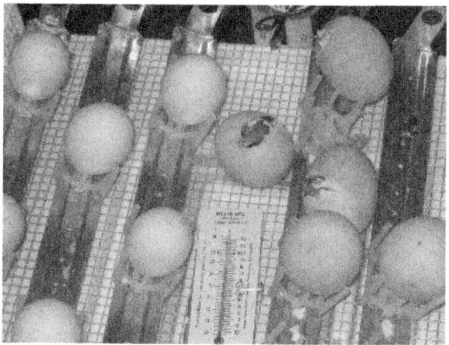

Once the egg has been pipped if, after 12 hours, the chick hasn't made any progress in breaking out, then you need to break only the top part of the shell away to help them. Don't break anymore. Some breeders feel you shouldn't help chicks that can't get out of the shell as you are breeding weakness into the bird, but others say you should as it can be due to a humidity problem or due to an inbred breed, such as exhibition strains.

Picture © Uberprutser

Once the chicks have hatched, leave them in the incubator until they have dried out fully before moving them to a heated, protected location. Newly hatched chicks do not need to eat for 24 hours after leaving the shell.

CARING FOR BABY CHICKS

Baby chicks are absolutely adorable; they are cute, fluffy and fun to watch. Children will fall in love with them and be fascinated by them. Baby chicks are also quite easy to handle which both kids and adults love.

Newly hatched chicks need a lot of care and monitoring, so you need about four weeks where you can give them a lot of attention. You will need to check on them five or more times each day, so no vacations when you've got baby chicks!

Before the birds hatch, you need to determine where you are going to keep them. As they are small they are easy to handle and can, at least initially, be put in a small space. Be aware though, that they will grow very quickly and soon take up a lot of space and make a lot of mess! Once they hit about five weeks old, you can move the birds into their outside coop, which makes your life much easier.

You need a predator and draft proof environment that doesn't get too cold, not your main living space, where you can keep your baby chicks. As they grow, they will kick up dust from scratching at their bedding, and they have a peculiar smell, which you probably don't want in your house!

You will need about two square feet per chick, which sounds like a lot when they are a day old, but they will soon grow much larger. They need protection from drafts, which can be in the form of a cardboard box with ventilation holes, a kids paddling pool or plastic storage bins.

Newly hatched chicks need a lot of heat, requiring an air temperature of 95F/35C in their first week, 90F/32C in their second week and then reducing a further 5F every week until they are ready to go outside.

The best heating system is a 250W infrared heat lamp suspended off the ground. How high the light is will depend on where it needs to be to achieve the right temperature, but you need to ensure that the chicks cannot touch the light and injure themselves. Do not use regular household light bulbs as they do not produce enough heat nor do they produce the right color of light, which often results in the chicks dying.

The red light is darker, meaning it doesn't disturb the chicks sleeping patterns and the color has also been found to help prevent chicks pecking each other.

Keep an eye on your chicks as their behavior can give you a good idea whether they have the correct conditions. If you see them crowded under the light then you know they are too cold, whereas if they are squashed against the outside of the brooder, they are too hot. A comfortable flock will be exploring their entire brooder.

Your baby chicks will need a number of items in their first home, including:

- Absorbent bedding – it is always a shock how much of a mess these little chicks can produce, but they do! Cover the floor of their home with pine shavings to a depth of 1". Avoid newspaper as it is not very absorbent and as it gets slippery can cause a deformity known as "splayed leg" which often results in other chickens pecking the deformed bird to death. Avoid cedar shavings as the oils in cedar affect the respiratory system later in life.
- Waterer – Don't use a dish or bowl for another animal as chicks can very easily fall into it and drown, spill it, walk in it, poop in it or kick their bedding into it. Avoid rabbit drinkers as it limits access to water to one chick at a time. Get a chick waterer as these prevent pollution of the water and allows multiple chicks to drink at once. In week two, raise it off the ground a little to help prevent contamination, but be prepared to change the water a few times each day.
- Feeder – like the waterer, do not use any old dish because it will get spilled and pooped on. Buy a genuine chick feeder such as that shown here. This saves you a lot of trouble and ensures the birds have easy access to their food.

- Roosting Poles – even baby chicks enjoy roosting, so provide them with some half inch diameter wooden dowel rods raised about 4 or

5 inches off the ground. This will help stop them trying to sit on the food and water containers!
- Feed – this comes in two forms, both specially prepared for chicks. The starter feed is either a crumb or mash, depend on how ground up it is. If you have had you chicks vaccinated then put them on an unmedicated feed, otherwise give them a medicated feed which is a great way to keep them healthy, but some people prefer organic feeds. Check the label on the food you buy as to when you switch the chicks to a growers feed, which is usually around 4 to 6 weeks, depending on the feed. Chicks will like a small amount of greens, bugs, and worms, though it is given to them as a treat after they have eaten their main food. Remember not to ration the food, and give your chicks all the food they want, they are quite capable of self-regulating how much they eat.
- Grit – chicks need grit, just like adult birds. You can use sand, canary gravel or parakeet gravel which can be found in local stores. Either sprinkle it on their food or give it to them in a separate bowl.
- Netting – unlike the adult birds, baby chicks are surprisingly good flyers and accomplished escape artists. Use some chicken wire or deer netting and drape it over the top of your brooder to stop the babies escaping before you are ready for them.

Baby chicks don't need a huge amount of equipment but keep them happy and content, and they won't fight. The time will fly by, and then they will be ready to go to the outside coop.

Chick Health Issues

Chicks do suffer from a number of health complaints which you need to keep an eye out for. In most cases, you will not see any of these, but you should be aware of them just in case they happen.

Pasting Up
This is where their droppings cake up on the vent and block it. It is very obvious and easy to spot but is deadly.

You must deal with this immediately using a warm, wet, soft paper towel which softens the poop. It then becomes easy to remove. Be careful to keep your chick warm during this process as the drying vent can make them cold.

In the worst cases, you must dunk the entire rear end of the chicken into warm water or even carefully use tweezers to remove the poop. If you do this then use a hair dryer on a gentle heat setting to dry the chick off, then wash your hands thoroughly.

Keep a close eye on the birds as this can come back again. This condition is more common in chicks that are sent through the mail rather than home hatched chicks.

Immediate Water Access
Although chicks don't need food in the first 24 hours, they do need water. Particularly when you order chicks through the mail, you must immediately get them to drink. This can involve taking the most active chick and gently dunk its beak into the water bowl. The other chicks should follow the lead to find the water. Do not feed your chicks water with a syringe as you can very easily drown them!

Umbilical Cord
Some chicks will have the umbilical cord still attached to its rear, which looks like a thin, black string. If you see this, then leave it alone. It will naturally fall off in a couple of days and pulling at it can seriously injure your chick!

Chick Checks
Check your chicks a few times a day. They'll be getting into trouble like kicking bedding around, throwing food around and all sorts. Watch for ones that are less active, keeping away from the rest of the flock or show other signs of something not being right. You are likely going to have to change their water and even their food a few times a day.

Changing Bedding
Baby chick bedding needs to be changed once or twice a week, depending on how many chicks you have and how messy they are. The used bedding can be composted.

Heating Requirements
Baby chicks need heat, to start with which is reduced as they grow older. In the first week, the temperature should be 95F/35C, the second week 90F/32C, reducing by five degrees each week. However, watch the chick's behavior to determine if you have the heating correct. If they huddle under the lamp, then they need more heat, whereas should they avoid the lamp then they are too hot.

Children and Chicks
It is very important that if children are going to interact with chicks that you

set boundaries. Baby chicks are delicate and will not appreciate being chased around or harassed. Loud noises will frighten them and cause stress.

Teach the children to be gentle and careful around chicks, showing them how to pick them up without hurting them. Any chicks which are not feeling well should be left alone by the children.

Kids will love chicks, but you have to keep a close eye on them, particularly younger ones as they can be excitable and over-active. It is good for them to interact and gets them involved in the hen keeping.

Outside Time

Once the chicks reach 3 or 4 weeks old, they can go outside in a protected run. It needs to be sunny with the temperature a minimum of 70F/21C. Make sure they have access to both water and shade and are well protected from potential predators. Also, ensure there is a cover over the run as the birds are good flyers and will escape the run. They will love digging around, but you must keep a close eye on them in case of predators or escape. If it gets windy, they will get cold, which they will let you know by their loud chirping!

At the age of 4 to 5 weeks your chicks will be ready to be moved outside all the time. They will love this, and the amount of care you need to give them will also be reduced. Make sure the outside environment is ready for them before this so you can easily move them with minimal disruption.

Continue to visit them a few times a day to keep an eye on them for the first week. Once they have settled, you can reduce your visits to a couple of times a day.

Holidaying When You Have Chickens

Owning chickens is a big commitment as they need regular care. When you are away from home, whether on holiday or business, you need to arrange for someone responsible for looking after your birds.

At a minimum, the chickens will need someone to let the chickens out of the coop in the morning and put them to bed at night. They will also need to replenish food and water and clean the coop and run. It is possible to get automatic door openers, but even so, your chickens need checking at least once every day.

Picture © Josh Larios

Make contact with local chicken owners and arrange with them to look after your chickens in return for you looking after theirs when they need it. Perhaps there is someone you know locally who will look after your hens in exchange for some eggs throughout the year.

As a last resort, you may have to hire someone such as a pet sitter or other local person. Ask around at local pet stores, feed stores and even vet's offices or look online. Whoever you choose, check their references and make sure they know exactly what they need to do. I'd recommend leaving written instructions and your phone number in case there are any problems.

Chickens don't like to be moved so ideally, you want to keep them at home in their own coop and run. If you have no choice, then you can relocate your chickens while you go away. Perhaps you have friends or family or know another hen keeper, but they live too far away for regular visits. In this case, you can move your chickens to their home, ensure they are secure and allow them to look after them. Just make sure your hens are safe from predators where they are staying and be away that stress from moving may cause them to be unsettled for a few days. This is best to do only when you are going away for longer periods of time.

There are people who offer boarding for chickens while their owners go on holiday. If you find one, you should ask them what they need from you. Some will require you to bring your own coop. Ensure that your hens will be placed on fresh ground and that everything is disinfected to avoid any chance of infection. You should visit the place before booking your chickens in to ensure it somewhere that they will be well looked after.

When boarding your chickens, send them with food to avoid any upset stomachs. Some boarding kennels will even provide nail trimming and wing clipping, so it is definitely worth looking at.

When you find someone, who is reliable and will look after your chickens you need to make their life as easy as possible. Leave things out where they are easy to find as well as plenty of food and bedding. They don't want to have to run to the store for supplies while you are away.

Before you leave, give everything a thorough clean and make sure everything is working and in good repair. Write a list of instructions for your chicken sitter so they know exactly what needs doing and make sure they can get in contact with you in case of emergency.

Moving Chickens

Chickens are not keen on being moved around. It causes them a lot of stress, which can make a weak bird go downhill. It can also stop their laying for a little while too.

When you move chickens adding a little apple cider vinegar to their water for a few days after the move helps their system cope with the stress. Dilute apple cider vinegar at 2% (2ml per liter of water) for adult chickens or 0.5% (5ml per liter of water) for chicks and growers.

When they are being transported, make sure they are comfortable and not over crowded. If they get too hot on the journey, they will get stressed. Transport them on the back seat of your car where you can monitor the temperature. If the weather is hot, then travel early in the morning or in the evening to prevent them overheating in the car. You should also never leave your chickens in the car as they could get too hot and die.

Chickens for Exhibitions & Shows

Although a lot of people will keep chickens for eggs or meat, some people like to keep them to exhibit at shows. It is a popular hobby, and you will need pure breeds that are recognized by the Poultry Society in your country. There are often local shows, county shows, state shows and even national shows, so there are plenty of events to attend.

Picture © Ben Stephenson

Shows are always run with very strict guidelines, and judges are typically experts in a breed or qualified to club standards. Before entering you need to familiarize yourself with the rules and ensure that both you and your chicken adhere to the rules. The shows will not tolerate any deviance from their regulation.

Typically, you are looking for a purebred chicken that is in good health and with good plumage. The coloring must conform to the breed standard otherwise it will be disqualified and it must conform to the right shape too.

Pure breed chickens are usually obtained from other breeders and will cost a premium. If you are serious about entering and winning competitions, you can obtain chicks from show winning chickens as they have a greater chance of inheriting the winning characteristics of their parents. However, these are going to be rare, may involve travel to collect and will be expensive.

Choosing a Breed
What breed to keep is a good question. Look at shows in your area and determine which breeds are often exhibited and which have categories that have a low number of entries. Talk to some of the exhibitors at a show and learn what their breeds need as well as any disadvantages of them. Remember that if you keep regular chickens, your show hens will require a separate coop and run to ensure they remain at show quality.

It is probably a good idea to avoid the rarer breeds initially as they can be harder to obtain and more difficult to look after.

Picture © Sergei Dmitriev

To start with, look for a breed that is a good layer and a good mother. This means you can easily breed your own chickens should you choose to do so. Your show chickens will need to be kept separate from any other chickens that you own.

Don't try to keep too many different varieties. Start off with one, maybe two at the most until you have experience in caring for show winning chickens.

If you are planning to breed your chickens, then you should make sure you keep very detailed breeding records to avoid breeding in deformities or unwanted traits. It is important that pure breed chickens adhere to the breed standard, so use leg bands to track your chickens during breeding. Before you start though, get a copy of the breed standard from your national poultry club and determine what the desirable characteristics are.

When your show birds grow up, you may need to separate them out, particularly the male birds as they can fight. If the birds do fight, it can cause damage to their comb, wattle or skin which will then make them undesirable for shows. Birds with fancy head feathers are probably also worth separating because they can get picked on because of their unusual feathers.

Once you know which breed(s) you want, then you can start to prepare them for the show.

Show Preparation

If you want to do well at a show, then you need to be prepared. The birds need to be well fed, relaxed and in good condition. Prepare well, and your chicken could end up being a champion!

Firstly, chickens do not travel well and do not like change so if the show requires a great deal of travel, get there a few days early so your chickens can recover from the stress of the journey and are calm during the exhibition. If you can take your chickens out in the car regularly, it will help them get used to being transported which reduces the stress on show days.

You will develop your own techniques for preparing your birds as you gain experience, but here are some general tips and advice to help the new exhibitor.

Birds must be pen trained, so they are relaxed throughout the exhibition. A bird sulking in its cage will tighten its feathers, peck at itself or kick bedding everywhere, particularly when the judge tries to get it out of the cage to look at it. The chicken should also be comfortable being handled, so they don't cause embarrassment when the judge examines them.

You can pen train your bird by putting it into a small pen, like those found at shows, handling it and giving it food so that it is calm and relaxed. The judges will use a stick to direct the chicken so you should get something similar and train the bird to display itself when directed by the stick. Talking to the birds as you work with them seems to have a very calming effect.

Before the show, there are a few things you can do to help your chickens look more presentable:

- Check the legs are not dry and scaly – if they are then use Vaseline, baby oil or olive oil to moisturize them. Old scales can sometimes be rubbed off carefully with your thumb nail.
- Remove dirt from under the legs scales carefully with a tooth pick.
- Trim beaks and nails so they are tidy using nail clippers but be very careful not to cut blood nerves.

Anywhere from a week to three days prior to the show you should wash your birds. This is particularly important with birds that have light colored feathers. Washing birds isn't difficult but will take a while to get used to.

You will need three baths containing warm water. The first has a pure soap or hair shampoo in. The other two are for rinsing the bird. If the bird is particularly dirty, then you can scrub it in the first tub, but make sure you scrub in the direction of the feathers, not against them.

Once the bird is clean, then rinse it in the second tub, pouring water on the bird to get all the soap off it and then rinse it again in the final tub.

Use a towel to soak excess water from the bird and then use a hair drier. Don't dry the bird fully as this can twist the feathers. Dry it around 80% or so, then let it either dry in the sun if it is warm enough or in front of a heater, so long as they can't get close enough to hurt themselves. They can very easily singe their feathers, which then means they won't do well in the show.

Dip a rag in either baby or olive oil and rub it on the comb, face, legs and ear lobes in order to enhance the brightness of the bird. Do this either the night before or the morning of the show. Avoid using too much oil as the bird will look too oily and be marked down.

Enhancing the bird's natural qualities is perfectly permittable in a show, but faking it is not just frowned upon, but is considered a serious issue and will result in instant disqualification and even banning from future shows. This includes coloring the feathers or ear lobes and even replacing feathers. The judges have years of experience and will easily spot this, so don't think you can fool them.

Taking part in a show involves completing an entry form which is returned to the show organizers a few weeks before the show. You will receive a schedule for the show that will tell you what is happening when it is happening and where it is happening. Pay attention to this so on show day you are in the right place at the right time.

Arrive at the show in plenty of time to put your birds in their pen and calm them down. Use strong cardboard boxes with ventilation holes and dry litter to transport the birds to the venue. The litter will soak up any droppings. However, your bird may still stand in them and make a mess of their feet, in which case wipe it off with a damp rag.

Once your birds are penned, you can relax, socialize and enjoy the show! It is always worth talking to the judges to get feedback about your chickens. You may get some tips and advice that will help you improve next time.

Exhibiting chickens is good fun, and some people will want to take their birds to shows. It is a fun hobby and very different from just keeping chickens for eggs or meat.

Making Money with Chickens

One of the perks of owning chickens is that they produce eggs, and depending on how many chickens you have and how many eggs they produce, you may have an excess to sell. It is unlikely that this will cover the cost of the chickens, but it can be a handy bit of extra income and help to reduce the cost of owning chickens. Selling meat from your chickens is a very different matter and subject to very strict regulations which means it is almost impossible for you to do. If you are interested in this, then check your local laws to determine whether it is possible.

Whether you can sell your eggs depends on the laws where you are. In the UK, for example, you are allowed to sell your eggs provided you have 49 or fewer hens. After you hit 50 hens, which most people won't, you must register with the Animal and Plant Health Agency. In the UK with less than 50 birds, you can sell your eggs without any registration or licenses as you are not considered a commercial egg producer. The exact rules may vary where you live, so check in to this before you start selling your eggs.

As a small time producer, you will be considered a Farm Gate Seller or something similar, which basically means you can sell your eggs from your home or even door to door in your area. Often, and certainly in the UK, you can sell your eggs directly to consumers at local markets, but your name and address must be on the box as must be advice on keeping eggs and a best before date (no more than 28 days after laying).

You will see commercially produced eggs marked with a stamp. As a Farm Gate Seller, you are exempt from this need, but you cannot sell your eggs as being graded in any way.

You will not be able to sell to shops, cafes, restaurants, hotels or anywhere else that isn't direct to the consumer. However, if you own a small B&B, you are usually allowed to serve your own eggs to your customers providing you inform them that they are from your chickens and that they are well cooked.

If you are selling eggs, then you need to ensure all your chickens are healthy. Sick birds need to be removed from the flock and its eggs discarded. If your chickens are taking medication, then check with the vet whether you can eat the eggs. Some medications will get into the eggs, making them unfit for human consumption.

Avoid selling any eggs that are cracked or damaged as once the shell is damaged harmful bacteria, or fungal spores can penetrate the egg.

Make sure the eggs you sell are fresh so as they are laid, put them into a box immediately and write the date on it. This ensures the eggs aren't damaged and that you know exactly when they should be used by.

Dirty eggs are not going to sell as people don't want eggs with chicken poop on. You can gently rub the eggs clean with a dry kitchen sponge, but don't use a wet cloth as it will remove the antibacterial protection (bloom) from the eggs. If you have to wash eggs, then wash them with warm water and then use them yourself rather than sell them.

Store the eggs before selling somewhere cool and dry, but do not refrigerate them. Guidelines state that eggs should not be refrigerated before sale because the temperature changes can cause condensation on the shell which then leads to fungal or bacterial damage.

Packaging Eggs for Sale
Eggs are best sold in egg boxes which protects them from damage. For

selling to friends and family, you can use egg boxes that you (or them) have saved from supermarket eggs.

If you are selling to the public, then you should use new boxes because it can be confusing for them to have two use by dates or other information on the box. You can also easily stick your name and address on to new boxes, and they look neat and tidy.

New cardboard egg boxes can be bought online. They are not expensive to buy and available either in clear plastic, cardboard or polystyrene. Some are even printed, whereas others are just plain.

Sizing Eggs
Supermarket eggs are sold according to size and all the eggs in a box will conform to that sizing standard. Fresh eggs come in all different sizes, depending on the breed of the bird, its age and more. The shell color can vary depending on the breed of hen you own, but be aware that unusual shell colorings can put some buyers off even though there is nothing wrong with your eggs.

In most cases, size isn't going to matter too much when selling fresh eggs, but you may find some people have supermarket led preconceptions of eggs which means they feel smaller eggs have something wrong with them. Stress that your eggs are fresh, local and natural plus they have a much better taste. Often the smaller eggs have larger yolks, which a lot of people will appreciate.

Some sellers will either try to group their eggs together by size or ensure there is a range of sizes, so each box contains roughly the same weight of eggs. If you have regular, repeat customers than try to make sure that the difference each week in egg size isn't too extreme.

Advertising Your Eggs
Marketing your eggs can be tricky as you need to get local buyers. A website or professional marketing materials certainly aren't worth the money due to your target market and low volume of sales.

Start by letting your friends and family know you are selling eggs and get them to pass the word about. Word of mouth is by far the best method of advertising and will get you started.

Depending on where you are located and zoning regulations, you could put a sign outside your house that you have eggs for sale. The downside of this is you will end up with complete strangers knocking at your door to buy eggs, which not everyone is happy about.

You can also put advertising flyers in local stores or community notice boards. In some areas, people leave their eggs by the road on a table with an honesty box, but this may not work in every area.

Local Facebook selling pages or sites such as Craigslist or Gumtree are also good places to advertising your eggs and allows you to make contact with potential buyers.

Legal Terminology
When describing your eggs, particularly in advertising, you need to be careful of the words you use as you could find yourself breaching advertising regulations.

You can advertise your eggs as fresh, because they are and also as non-cage eggs because your birds are not caged, they have a coop and a run.

In order to use the term Free Range, your chickens must have constant access during the day to an outside space that contains vegetation. How much space they need will depend on where you live, but in the UK, they need a minimum of 4 square meters of outside space per hen. For many backyard farmers, this just isn't possible, so you will be unable to describe your eggs as Free Range.

Organic is another term you must be very careful about using. To be organic hens need to be free range and have been fed on only a certified organic feed. There may well be other conditions attached to using this label, depending on where you live. In the UK hens are not allowed any antibiotics unless it is a medical emergency, must not have been fed any animal by-products or had any food from genetically modified plants.

Be very careful about using these labels unless you are completely sure you can do so. Incorrect usage can result in legal fees and fines, which are unlikely to be covered by what you will ever make selling eggs.

Selling eggs isn't going to make you a fortune. At the very best it will help get rid of an excess of eggs and maybe offset some of the cost of owning chickens. Look around your local area to see what price you can sell your eggs at. As they are fresh and natural, they usually command a premium, but be careful of pricing yourself out of the market. Check locally what other people are selling eggs at to determine the price the local market can handle. Almost everyone locally will be selling their chicken eggs at more or less the same price.

Chicken Keeping Tips

Keeping chickens has become very popular as increasing numbers of people want healthy, fresh eggs that are free from antibiotics and other chemicals. Lots of breeders are appearing all over the place that sell not only healthy chickens but provide advice and help to new keepers.

There are plenty of challenges associated with keeping chickens but, having read this book you should be well equipped to deal with almost all of them. These are some of the best tips I've gathered to help you avoid common mistakes and get the most from your chicken keeping hobby.

Have a Plan
Before you purchase your first hen, have a plan on what coop you are buying or making, the run you need and everything else. If you are building it yourself, get some plans and make sure you know exactly what you are doing and that your coop will be safe for your hens.

Find Other Chicken Keepers
There are bound to be other people in your local area that keep chickens so make contact with them. Not only are they a good source of advice and help, but they could also make great chicken sitters if you go on holiday! Seeing their set-ups will help you to determine what you need and what is involved in keeping chickens before you have spent any money.

You Will Lose Some Chickens
Sorry, but it is going to happen, no matter how good you are and how well you protect your hens. Whether it is a fox, dog, raccoon, weasel or disease, something, at some point, is going to get your chickens. Protect your hens as best you can and design your coop/run to be as predator proof as possible, but be prepared for the worst to happen one day.

Dump the Cockerels
Although a lot of people can keep chickens, cockerels are usually not permitted and are notoriously difficult to rehome. Suburban areas are not going to appreciate a rooster crowing at dawn every morning, so you need to get rid of the male birds. If you buy eggs to hatch, you can expect a little more than half to be male. If you buy day old chicks, it may be several weeks before you can sex the bird, unless it is an autosexing breed. If you got your chicks from a local person, then often they will take the males back and replace them with females.

Inconsistent Laying
Egg laying is a natural process and can be disturbed by all sorts of factors, many of which are out of your control. We had a cool, damp summer here this year which wreaked havoc on egg production. The amount of eggs you get will depend on the breed, so regularly check for eggs, use what you find and be patient with your birds.

Research and Learn
There is always so much to learn about owning chickens, and even a book like this cannot possibly communicate every iota of knowledge. Get involved in chicken Facebook groups and forums to find out more about chickens. Check out online blogs too. You will learn lots of useful tips plus find out what other people are doing. You'll also discover what the latest developments and techniques are in keeping chickens at home.

Enjoy Your Eggs
Freshly laid eggs have an amazing taste, much better than anything you can get from the supermarket. This is probably the main reason people keep chickens. Enjoy your chickens and their eggs; you'll be the envy of your

friends!

Plan for When Production Drops
Chickens have a finite egg producing life and there comes the point where they either stop laying completely, or they are no longer cost effective to keep, i.e. the cost of feed outweighs the saving from not buying eggs. At this point, you must either eat the bird (butchering it yourself or sending it to a butcher) or allow the bird to be a pet and die of old age. Bear in mind that this chicken will be taking up space that an egg laying chicken could have and eventually you will end up with a flock of pets and no eggs!

ENDNOTE

Keeping chickens is a fantastic hobby and very rewarding. Hens make surprisingly good pets, are entertaining to watch and you get the benefit of fresh eggs too!

This isn't something you should just leap in to, but you need to plan and prepare. Reading this book is a good first step, and now you are aware of what you need for keeping hens.

Start by working out where you can keep your chickens in your garden and how much space they will have. Once this is done, measure up coops and runs, and then you can determine exactly how many chickens you have the space to keep. Remember that they can live for a number of years so be prepared for a long-term commitment. Check with local zoning and planning laws as well as any HOA regulations as to whether you can keep chickens.

Before you even see a chicken, you need to get their home set up, predator proofed and buy all the equipment you need. This initial outlay will be quite significant, but afterward there are only feed and bedding costs plus occasional vet fees.

Predator proofing is absolutely vital as this is the main way you will lose members of your flock. Take the time to ensure the coop and run is predator proof before you introduce chickens to it as this work is much easier to do without chickens helping and escaping!

Choose your breed well based on the size bird you are after and the traits you want. Remember that some birds are good layers but poor meat birds, some are good all-rounders and others make great meat birds but produce few eggs. Avoid mixing breeds initially, and certainly, never mix more aggressive breeds with more timid breeds as they will fight.

Check your chickens daily for signs of ill health. Like many prey animals, they are very good at hiding signs of illness, but you will learn to spot them and take the required early action to save them.

Ensure your chickens get the right food and regular treats. It will keep them healthy and happy, though avoid feeding them the foods which are bad for them or even poisonous to them. Check their run and around their living quarters regularly for any poisonous plants growing and remove them immediately.

So long as you provide your chickens with enough space to run around and roost, good nesting boxes and plenty of food, water, and grit, they will provide you with fresh eggs. Egg production falls when the weather gets cooler and stops during molting, but for much of the rest of the year, you will get regular fresh eggs.

Whatever your plans for your chickens, whether eggs, meat or exhibiting, you will find that they are great animals to keep. They can be very sociable, and kids enjoy being involved in their upkeep. Be aware though that it is a significant time commitment as you need to visit your hens a minimum of twice a day, every day. This can make vacations difficult unless you can find a chicken sitter (try local keepers as they won't mind trading chicken sitting duties with you), but it is rewarding.

With more people being concerned about where their food comes from and what goes into their food, chickens are becoming very popular because you know exactly where your meat and eggs come from. Of course, butchering a chicken isn't for everyone, and a lot of people will not want this job. Luckily, you will be able to find a local butcher who will happily take care of your chicken for you, for a fee, and return you a neatly packaged chicken that is ready to cook.

Enjoy keeping your chickens and remember that it is a constant learning process. It is great fun, and there are lots of rewards, but be prepared to commit to your chickens before you buy them. The fresh eggs alone make the process worthwhile, and sitting watching chickens is a great stress reliever as they are fun, entertaining birds!

ABOUT JASON

Jason has been a keen gardener for over twenty years, having taken on numerous weed infested patches and turned them into productive vegetable gardens.

One of his first gardening experiences was digging over a 400 square foot garden in its entirety and turning it into a vegetable garden, much to the delight of his neighbors who all got free vegetables! It was through this experience that he discovered his love of gardening and started to learn more and more about the subject.

His first encounter with a greenhouse resulted in a tomato infested greenhouse but he soon learnt how to make the most of a greenhouse and now grows a wide variety of plants from grapes to squashes to tomatoes and more. Of course, his wife is delighted with his greenhouse as it means the windowsills in the house are no longer filled with seed trays every spring.

He is passionate about helping people learn to grow their own fresh produce and enjoy the many benefits that come with it, from the exercise of gardening to the nutrition of freshly picked produce. He often says that when you've tasted a freshly picked tomato you'll never want to buy another one from a store again!

Jason is also very active in the personal development community, having written books on self-help, including subjects such as motivation and confidence. He has also recorded over 80 hypnosis programs, being a fully qualified clinical hypnotist which he sells from his website www.MusicForChange.com.

He hopes that this book has been a pleasure for you to read and that you have learned a lot about the subject and welcomes your feedback either directly or through an Amazon review. This feedback is used to improve his books and provide better quality information for his readers.

Jason also loves to grow giant and unusual vegetables and is still planning on breaking the 400lb barrier with a giant pumpkin. He hopes that with his new allotment plot he'll be able to grow even more exciting vegetables to share with his readers.

Other Books By Jason

Please check out my other gardening books on Amazon, available in Kindle and paperback.

Berry Gardening – The Complete Guide to Berry Gardening from Gooseberries to Boysenberries and More
Who doesn't love fresh berries? Find out how you can grow many of the popular berries at home such as marionberries and blackberries and some of the more unusual like honeyberries and goji berries. A step by step guide to growing your own berries including pruning, propagating and more. Discover how you can grow a wide variety of berries at home in your garden or on your balcony.

Canning and Preserving at Home – A Complete Guide to Canning, Preserving and Storing Your Produce
A complete guide to storing your home-grown fruits and vegetables. Learn everything from how to freeze your produce to canning, making jams, jellies, and chutneys to dehydrating and more. Everything you need to know about storing your fresh produce, including some unusual methods of storage, some of which will encourage children to eat fresh fruit!

Companion Planting Secrets – Organic Gardening to Deter Pests and Increase Yield
Learn the secrets of natural and organic pest control with companion planting. This is a great way to increase your yield, produce better quality plants and work in harmony with nature. By attracting beneficial insects to your garden, you can naturally keep down harmful pests and reduce the damage they cause. You probably grow many of these companion plants already, but by repositioning them, you can reap the many benefits of this natural method of gardening.

Container Gardening - Growing Vegetables, Herbs & Flowers in Containers

A step by step guide showing you how to create your very own container garden. Whether you have no garden, little space or you want to grow specific plants, this book guides you through everything you need to know about planting a container garden from the different types of pots, to which plants thrive in containers to handy tips helping you avoid the common mistakes people make with containers.

Cooking with Zucchini - Delicious Recipes, Preserves and More With Courgettes: How To Deal With A Glut Of Zucchini And Love It!

Getting too many zucchinis from your plants? This book teaches you how to grow your own courgettes at home as well as showing you the many different varieties you could grow. Packed full of delicious recipes, you will learn everything from the famous zucchini chocolate cake to delicious main courses, snacks, and Paleo diet friendly raw recipes. The must have guide for anyone dealing with a glut of zucchini.

Environmentally Friendly Gardening – Your Guide to a Sustainable, Eco-Friendly Garden

With a looming environmental crisis, we are all looking to do our bit to save the environment. This book talks you through how to garden in harmony with nature and reduce your environmental impact. Learn how to eliminate the need for chemicals with clever techniques and eco-friendly alternatives. Discover today how you can become a more environmentally friendly gardener and still have a beautiful garden.

Greenhouse Gardening - A Beginners Guide to Growing Fruit and Vegetables All Year Round

A complete, step by step guide to owning your own greenhouse. Learn everything you need to know from sourcing greenhouses to building foundations to ensuring it survives high winds. This handy guide will teach you everything you need to know to grow a wide range of plants in your greenhouse, including tomatoes, chilies, squashes, zucchini and much more. Find out how you can benefit from a greenhouse today, they are more fun and less work than you might think!

Growing Brassicas – Growing Cruciferous Vegetables from Broccoli to Mooli to Wasabi and More

Brassicas are renowned for their health benefits and are packed full of vitamins. They are easy to grow at home but beset by problems. Find out how you can grow these amazing vegetables at home, including the incredibly beneficial plants broccoli and maca. Includes step by step growing guides plus delicious recipes for every recipe!

Growing Chilies – A Beginners Guide to Growing, Using & Surviving Chilies

Ever wanted to grow super-hot chilies? Or maybe you just want to grow your own chilies to add some flavor to your food? This book is your complete, step-by-step guide to growing chilies at home. With topics from selecting varieties to how to germinate seeds, you will learn everything you need to know to grow chilies successfully, even the notoriously difficult to grow varieties such as Carolina Reaper. With recipes for sauces, meals and making your own chili powder, you'll find everything you need to know to grow your own chili plants

Growing Fruit: The Complete Guide to Growing Fruit at Home

This is a complete guide to growing fruit from apricots to walnuts and everything in between. You will learn how to choose fruit plants, how to grow and care for them, how to store and preserve the fruit and much more. With recipes, advice, and tips this is the perfect book for anyone who wants to learn more about growing fruit at home, whether beginner or experienced gardener.

Growing Garlic – A Complete Guide to Growing, Harvesting & Using Garlic

Everything you need to know to grow this popular plant. Whether you are growing normal garlic or elephant garlic for cooking or health, you will find this book contains all the information you need. Traditionally a difficult crop to grow with a long growing season, you'll learn the exact conditions garlic needs, how to avoid the common problems people encounter and how to store your garlic for use all year round. A complete, step-by-step guide showing you precisely how to grow garlic at home.

Growing Herbs – A Beginners Guide to Growing, Using, Harvesting and Storing Herbs

A comprehensive guide to growing herbs at home, detailing 49 different herbs. Learn everything you need to know to grow these herbs from their preferred soil conditions to how to harvest and propagate them and more. Including recipes for health and beauty plus delicious dishes to make in your kitchen. This step-by-step guide is designed to teach you all about growing herbs at home, from a few herbs in containers to a fully-fledged herb garden. An indispensable guide to growing and using herbs.

Growing Giant Pumpkins – How to Grow Massive Pumpkins at Home

A complete step by step guide detailing everything you need to know to produce pumpkins weighing hundreds of pounds, if not edging into the thousands! Anyone can grow giant pumpkins at home, and this book gives you the insider secrets of the giant pumpkin growers showing you how to avoid the mistakes people commonly make when trying to grow a giant pumpkin. This is a complete guide detailing everything from preparing the soil to getting the right seeds to germinating the seeds and caring for your pumpkins.

Growing Lavender: Growing, Using, Cooking and Healing with Lavender

A complete guide to growing and using this beautiful plant. Find out about the hundreds of different varieties of lavender and how you can grow this bee friendly plant at home. With hundreds of uses in crafts, cooking and healing, this plant has a long history of association with humans. Discover today how you can grow lavender at home and enjoy this amazing herb.

Growing Tomatoes: Your Guide to Growing Delicious Tomatoes at Home

This is the definitive guide to growing delicious and fresh tomatoes at home. Teaching you everything from selecting seeds to planting and caring for your tomatoes as well as diagnosing problems this is the ideal book for anyone who wants to grow tomatoes at home. A comprehensive must have guide.

How to Compost – Turn Your Waste into Brown Gold

This is a complete step by step guide to making your own compost at home. Vital to any gardener, this book will explain everything from setting up your compost heap to how to ensure you get fresh compost in just a few weeks. A must have handbook for any gardener who wants their plants to benefit from home-made compost.

How to Grow Potatoes - The Guide to Choosing, Planting and Growing in Containers Or the Ground

Learn everything you need to know about growing potatoes at home. Discover the wide variety of potatoes you can grow, many delicious varieties you will never see in the shops. Find out the best way to grow potatoes at home, how to protect your plants from the many pests and diseases and how to store your harvest so you can enjoy fresh potatoes over winter. A complete step by step guide telling you everything you need to know to grow potatoes at home successfully.

Hydroponics: A Beginners Guide to Growing Food without Soil
Hydroponics is growing plants without soil, which is a fantastic idea for indoor gardens. It is surprisingly easy to set up, once you know what you are doing and is significantly more productive and quicker than growing in soil. This book will tell you everything you need to know to get started growing flowers, vegetables, and fruit hydroponically at home.

Indoor Gardening for Beginners: The Complete Guide to Growing Herbs, Flowers, Vegetables and Fruits in Your House
Discover how you can grow a wide variety of plants in your home. Whether you want to grow herbs for cooking, vegetables or a decorative plant display, this book tells you everything you need to know. Learn which plants to keep in your home to purify the air and remove harmful chemicals and how to successfully grow plants from cacti to flowers to carnivorous plants.

Raised Bed Gardening – A Guide to Growing Vegetables In Raised Beds
Learn why raised beds are such an efficient and effortless way to garden as you discover the benefits of no-dig gardening, denser planting and less bending, ideal for anyone who hates weeding or suffers from back pain. You will learn everything you need to know to build your own raised beds, plant them and ensure they are highly productive.

Save Our Bees – Your Guide to Creating a Bee Friendly Environment
Discover the plight of our bees and why they desperately need all of our help. Find out all about the different bees, how they are harmless, yet a vital part of our food chain. This book teaches you all about bees and how you can create a bee friendly environment in your neighborhood. You will learn the plants bees love, where they need to live and what plants are dangerous for bees, plus lots, lots more.

Vertical Gardening: Maximum Productivity, Minimum Space
This is an exciting form of gardening allows you to grow large amounts of fruit and vegetables in small areas, maximizing your use of space. Whether you have a large garden, an allotment or just a small balcony, you will be able to grow more delicious fresh produce. Find out how I grew over 70 strawberry plants in just three feet of ground space and more in this detailed guide.

Worm Farming: Creating Compost at Home with Vermiculture
Learn about this amazing way of producing high-quality compost at home by recycling your kitchen waste. Worms break it down and produce a sought after, highly nutritious compost that your plants will thrive in. No matter how big your garden you will be able to create your own worm farm and compost using the techniques in this step-by-step guide. Learn how to start worm farming and producing your own high-quality compost at home.

Want More Inspiring Gardening Ideas?

This book is part of the Inspiring Gardening Ideas series. Bringing you the best books anywhere on how to get the most from your garden or allotment. Please remember to leave a review on Amazon once you have finished this book as it helps me continually improve my books.

You can find out about more wonderful books just like this one at: www.GardeningWithJason.com

Follow me at www.YouTube.com/OwningAnAllotment for my video diary and tips. Join me on Facebook for regular updates and discussions at www.Facebook.com/OwningAnAllotment.

Find me on Instagram and Twitter as @allotmentowner where I post regular updates, offers and gardening news. Follow me today and let's catch up in person!

Free Book!

Visit http://gardeningwithjason.com/your-free-book/ now for your free copy of my book "Environmentally Friendly Gardening" sent to your inbox. Discover today how you can become a more eco-friendly gardener and help us all make the world a better place.

This book is full of tips and advice, helping you to reduce your need for chemicals and work in harmony with nature to improve the environment. With the looming crisis, there is something we can all do in our gardens, no matter how big or small they are and they can still look fantastic!

Thank you for reading!

Printed in Great Britain
by Amazon